Engineering Materials

This series provides topical information on innovative, structural and functional materials and composites with applications in optical, electrical, mechanical, civil, aeronautical, medical, bio- and nano-engineering. The individual volumes are complete, comprehensive monographs covering the structure, properties, manufacturing process and applications of these materials. This multidisciplinary series is devoted to professionals, students and all those interested in the latest developments in the Materials Science field, that look for a carefully selected collection of high quality review articles on their respective field of expertise.

Indexed at Compendex (2021) and Scopus (2022)

Arnold C. Alguno · Rey Y. Capangpangan ·
Gerard G. Dumancas · Arnold A. Lubguban ·
Roberto M. Malaluan · Rolen Brian P. Rivera

Metal Nanoparticles-Based Nanoplatforms for Colorimetric Sensing

Mechanisms, Applications, and Future Directions

Springer

Arnold C. Alguno
Department of Materials and Resources Engineering and Technology
Mindanao State University—Iligan Institute of Technology
Iligan City, Philippines

Rey Y. Capangpangan
Department of Physical Sciences and Mathematics, College of Marine and Allied Sciences
Mindanao State University at Naawan
Naawan, Philippines

Gerard G. Dumancas
Honors College and the Department of Chemistry
North Carolina Agricultural and Technical State University
Greensboro, NC, USA

Arnold A. Lubguban
Department of Chemical Engineering and Technology
Mindanao State University—Iligan Institute of Technology
Iligan City, Philippines

Roberto M. Malaluan
Department of Chemical Engineering and Technology
Mindanao State University—Iligan Institute of Technology
Iligan City, Philippines

Rolen Brian P. Rivera
Mindanao State University—Iligan Institute of Technology
Iligan City, Philippines

ISSN 1612-1317 ISSN 1868-1212 (electronic)
Engineering Materials
ISBN 978-981-96-6770-3 ISBN 978-981-96-6771-0 (eBook)
https://doi.org/10.1007/978-981-96-6771-0

This Springer imprint is published by the registered company Springer Nature Singapore Pte Ltd.
The registered company address is: 152 Beach Road, #21-01/04 Gateway East, Singapore 189721, Singapore

Preface

The effectiveness of nanotechnology represents a paradigm shift in analytical chemistry, combining accuracy, simplicity, and high sensitivity as a solution to the greatest challenges in environmental monitoring, food safety, and medical diagnostics. This book, *Metal Nanoparticles-Based Nanoplatforms for Colorimetric Sensing: Mechanisms, Applications, and Future Directions*, compiles pioneering research and innovation in this rapidly evolving field. This review highlights the excellent properties of metal nanoparticles (MNPs) such as gold, silver, and copper, in forming nanoplatforms that exploit their optical and catalytic properties for highly sensitive colorimetric detection.

The effectiveness of metal nanoparticles is partly attributed to their ability to exhibit localized surface plasmon resonance, which produces visible color changes indicating molecular interactions in solution. This deceptively simple yet scientifically rich observation has led to rapid advances in the detection of a wide variety of analytes, from toxic heavy metals to clinically important biomarkers. By leveraging these properties, the research community has developed durable and easy-to-use sensing devices that enable real-time diagnostics on-site that would otherwise be difficult to achieve.

Theoretical concepts, experimental methodologies, and practical applications are consolidated in this book, making it a significant reference for scientists, engineers, and technologists. The chapters carefully examine the complex properties of metal nanoparticles, explore their cooperative effects in hybrid nanostructures, and address critical such as reproducibility, cost efficiency, and scalability. The book also discusses the future perspectives, including the integration of artificial intelligence and sustainable synthesis methods that are poised to reshape nanotechnology in the years ahead.

In tribute to this collaborative endeavor of scientific creativity, the current work encourages researchers to harness the capabilities of MNPs for colorimetric sensing, fostering innovations that promote sustainability, public health, and environmental stewardship. Readers are invited to explore this dynamic field of nanoscience, where discovery meets application and challenges evolve into opportunities.

Iligan City, Philippines	Arnold C. Alguno
Naawan, Philippines	Rey Y. Capangpangan
Greensboro, North Carolina, USA	Gerard G. Dumancas
Iligan City, Philippines	Arnold A. Lubguban
Iligan City, Philippines	Roberto M. Malaluan
Iligan City, Philippines	Rolen Brian P. Rivera

Competing Interests The authors have no competing interests to declare that are relevant to the content of this manuscript.

Contents

About the Authors

Arnold C. Alguno is Professor in the Department of Materials and Resources Engineering and Technology and Director for the Office of Research Management at Mindanao State University—Iligan Institute of Technology (MSU-IIT), Philippines. He works on semiconductor physics, solar cell technology, quantum dots, sensor technology, nanotechnology, and advanced functional materials. Dr. Alguno received his M.Sc. in physics at MSU-IIT and doctor of philosophy in physics and Post-Doctoral Fellow at Tohoku University in Japan. His research area primarily covers the synthesis and characterization of nanomaterials and polymeric materials, contributing to technological advancements in material science and nanotechnology.

Rey Y. Capangpangan is Professor of chemistry at Mindanao State University in Naawan (MSUN), Philippines. He is Distinguished Researcher and Educator specializing in catalysis for environmental applications and the development of advanced nano-based catalysts. Dr. Capangpangan has led impactful projects on synthesizing and characterizing catalyst materials aimed at environmental remediation, pollution control, and sustainable energy solutions. His groundbreaking work in environmental catalysis addresses critical challenges in water and air pollution, establishing him as a leading voice in the field of environmental chemistry. Dr. Capangpangan's research integrates chemistry, materials science, and engineering, fostering interdisciplinary approaches to sustainable development. His commitment to research, teaching, and community engagement underscores his influential role in advancing catalytic science for environmental sustainability.

Gerard G. Dumancas is Associate Dean of Research and Program Innovation (Honors College) and Professor of chemistry at North Carolina Agricultural and Technical State University. He was previously an Associate Professor of Chemistry and the Director of the National Science Foundation (NSF) Robert Noyce Teacher Scholarship Program at the University of Scranton in Scranton, P A. At the University of Scranton, he also served as a Faculty Fellow of the Office of Community-Based Learning, Office of First Year Residential Communities, and the Office of the Educational Assessment. He received his B.S. in chemistry from the University of the Philippines in 2005 and his Ph.D. in analytical chemistry (Chemometrics) from Oklahoma State University in 2012. He completed his postdoctoral training at the Oklahoma Medical Research Foundation in the area of statistical genetics and bioinformatics. His research has been recognized by the American Oil Chemists' Society, the Society for Laboratory Automation and Screening, the NSF Noyce Scholarship Program, the NSF S-STEM Program, and the Louisiana State University Leveraging Innovation for Technology Transfer (LSU LIFT^2) grant program. He was previously named as an NSF Program to Empower Partnerships with Industry Fellow as well as Visiting Scientist of the US Department of Energy's Joint Bioenergy Institute and the National Renewable Energy Laboratory, and the Berkeley Lab.

Arnold A. Lubguban currently holds an appointment as Professor and Senior Researcher at the Mindanao State University—Iligan Institute of Technology's (MSU-IIT) in the Philippines. Before joining MSU-IIT in 2015, he had Short-Term Balik-Scientist stint with the Philippines' Department of Science and Technology (DOST) in 2014, giving lectures and training on bio-based polymer processing to academe and local industries. He received his B.S. in 1999 at the University of the Philippines—Los Baños (UPLB) and the University of Missouri's M.S. and Ph.D. programs in chemical engineering in 2005 and 2009, respectively. He did a postdoc in industrial polymer processing in 2009 at the Ohio State University, focusing on biomass conversion into polyols and polyurethanes. He led several research programs and project leadership under the USAID-STRIDE and DOST-PCIEERD on processing sustainable polymers for industrial applications.

Roberto M. Malaluan was born in San Jose, Batangas, Philippines, and attended the Mindanao State University—Marawi, receiving a B.S. in 1983. He worked as Engineering Instructor at Mindanao State University—Iligan Institute of Technology (MSU-IIT) for several years before attending Tohoku University in Sendai, Japan, where he received his Ph.D. under the guidance of Kunio Arai in 1995. He became Full Professor of chemical engineering at MSU-IIT in 2004. He led several project leadership positions on supercritical carbon dioxide extraction of selected natural products and indigenous species for drug discovery and industrial applications under the Philippines' Department of Agriculture (DA) and the Department of Science and Technology (DOST). He works as Senior Faculty Member doing instruction and research on sustainable polymers.

Rolen Brian P. Rivera finished his doctor of philosophy in physics at Mindanao State University—Iligan Institute of Technology (MSU-IIT), Philippines. He was Research Fellow under the CHED-LAKAS-funded Colorimetric Project at MSU-IIT. His work focuses on developing a colorimetric sensor for detecting food spoilage. He holds both B.Sc. and M.Sc. degrees in physics from MSU-IIT, where he gained extensive expertise in synthesizing and characterizing metal nanoparticles for colorimetric applications. His research contributes to advancing accessible, rapid sensing technologies for food safety and quality control.

Chapter 1
Introduction to Metal Nanoparticle-Based Colorimetric Sensing

Abstract This chapter covers the progress of the colorimetric sensing based on metal nanoparticle, which is experiencing rapid development in analytical chemistry. Colorimetric sensors allow simple, on-site detection of chemical and biological analytes, based on the distinctive optical properties of gold (AuNPs) and silver (AgNPs) nanoparticles. These sensors work based on a localized surface plasmon resonance (LSPR) that causes a color change that can potentially be detected visually by the naked eye, thus increasing accessibility and visibility for non-specialists. Different strategies to enhance the sensitivity and selectivity of these sensors, such as the functionalization of nanoparticles with aptamers, ligands, and polymers provide an insight in this chapter. When coupled with various nanoplasmonic substrates, colorimetric sensing provides an interpretation of color change that can eventually read trace analyte amounts, thus providing new and better diagnostic methods for pollutants, pathogens, and biomarkers for environmental monitoring, food safety, and healthcare applications. In addition, this chapter tries to highlight the practicality of the nanoparticle synthesis processes and their applications in the form of rapid diagnostic kits and portable monitoring tools, providing real solutions to some of real world problems.

1.1 Overview of Colorimetric Sensing

Colorimetric sensing measures the change in color of the solution responding to specific stimuli which are very effective and selective in the detection of various chemical and biomolecules. This method utilizes different materials such as biopolymers, chromogenic compounds, plasmonic nanomaterials, and metal complexes. The process of detection depends on mechanisms such as via photo-induced electron transfer and sol–gel reactions [4, 13, 25, 29]. For example, metal complexes and chromogenic compounds are usually matched with electrochemical and piezoelectric materials. On the other hand, biopolymers are important in immuno- and gene sensors. Inorganic materials are used in optical sensing due to their high electron transport and superior optical properties [8, 23].

A. C. Alguno et al., *Metal Nanoparticles-Based Nanoplatforms for Colorimetric Sensing*, Engineering Materials, https://doi.org/10.1007/978-981-96-6771-0_1

Colorimetric sensing is commonly used in environmental monitoring, clinical diagnostics, and food safety [14, 28]. Color changes are detected in water flow utilizing cheap sensor chips with efficient results. Noble metal nanoparticles, such as gold and silver, enhance sensitivity due to their localized surface plasmon resonance [4, 10, 26]. Moreover, using aptamers improve the selectivity of these nanoparticles for effective sensing capability. This approach utilizes various materials and mechanisms that improve the detection of chemicals and biomolecules.

1.1.1 *Basic Principles and Historical Development of Colorimetric Detection*

Colorimetric detection can be observed via color changes in the solution. This color change is predominantly influenced by varying the concentration of the colored substance, which is primarily based on Beer-Lambert Law stating that as concentration increases, absorbance also rises. Good examples of this colorimetric detection are Nessler's reagent for ammonia concentration and litmus paper for pH measurement [24]. In the nineteenth century, it was observed that enzyme-based reagents were used to measure glucose in urine and blood alcohol levels. The innovation timeline of colorimetric sensing as applied analytically is shown in Fig. 1.1. The colorimetric sensing application has significantly evolved. It started from traditional paper-based methods and evolved to advanced microfluidic devices. Modern technological advancements offer opportunities for real-time rapid kinetic measurements and multispecies analysis which can be applied to technology and telecommunications [2]. Gold and silver noble metal nanoparticles are used to enhance sensitivity of analytes due to their localized surface plasmon resonance (LSPR) and adding aptamers to improve their selectivity and reactivity. This approach is used to efficiently detect various biomolecules and chemicals.

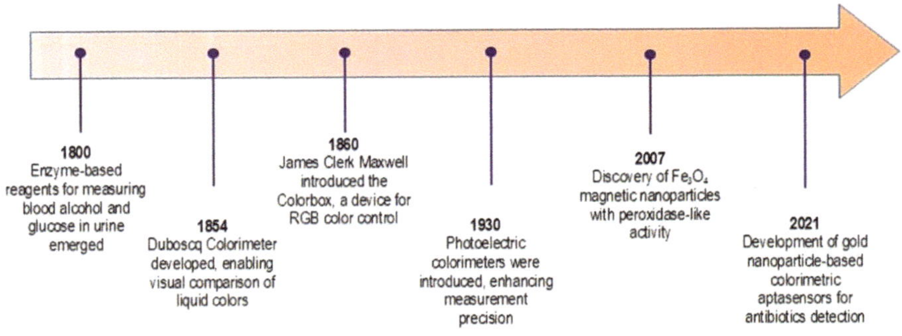

Fig. 1.1 Innovations timeline of colorimetric sensing as applied analytically: from early reagents to advanced nanoparticle sensor

1.1.2 Visual-Based Detection Methods in Analytical Chemistry

Visual-based detection methods are very important in analytical chemistry, particularly for colorimetric detection and assays in which color changes can be observed in the presence of an analyte [25]. The simplicity and label-free operation are the advantages of these methods for easy interpretation outside laboratories for on-site applications. Naked-eye colorimetric detection and chromogenic substrates are among these strategies. Additionally, the colorimetric detection sensitivity can be enhanced by utilizing metal nanoparticles that can efficiently detect trace targets. Furthermore, several improvements in this field emerged that include more complex devices for simultaneous multispecies analysis with enhanced sensitivity using noble metal nanoparticles.

The power of metal nanoparticles serve as mimetics or catalysts in converting trace amounts of long-wavelength chromogenic substrates to observe changes in color responses through covalent aggregation or localized surface plasmon resonance (LSPR). Visual detection methods are essential in environmental monitoring and biological chemistry. However, these methods are very subjective and affected by personal variations of visual observation which could result in complicated quantitative analysis. That is why many detection techniques are very dependent on instruments to evaluate sample color for quantitative concentration analysis. Smartphones are now used as color-sensing devices with improved subjective detection to monitor analytes in numerous applications.

Embracing visual-based methods from traditional colorimetric detection signifies a significant improvement in the field of analytical chemistry. Incorporation of nanotechnology to achieve enhanced sensitivity and accessibility experience a substantial improvement as compared to early techniques which simply rely on color changes. A wide range of applications including on-site monitoring to various complex laboratory assays utilize these visual-based techniques that offer a simplified detection process. Nowadays, accurate and user-friendly detection is the utmost consideration of end-use. Utilizing chromogenic substrates and metal nanoparticles could facilitate the merging of ancient practices with today's modern technology for enhanced colorimetric detection.

1.2 The Role of Metal Nanoparticles in Colorimetric Platforms

In early clinical disease diagnosis and environmental pollution screening, it is very important to detect various analytes in fluids, and, therefore methods like colorimetric sensing are essential [4, 25]. The unique optical properties of metal nanoparticles are useful to fabricate colorimetric platforms to clearly observe naked-eye color changes. The presence of organic compounds and biomolecules in a compound or analyte are factors that influence the color changes and this can also be affected by the amount of heat and intensity of light. Further, the design of colorimetric sensors based on metal nanoparticles will take into account the type of nanoparticles and their characteristic color responses. The recent advancements and fundamental aspects of the design are discussed in the succeeding sections.

The color changes phenomenon is influenced by the interaction between the analytes and metal nanoparticles that would eventually trigger self-assembly, aggregation, and rearrangement. Gold nanoparticles (AuNPs) and silver nanoparticles (AgNPs) are utilized in colorimetric sensing due to their strong absorption characteristic peaks associated with LSPR. In particular, AuNPs exhibit strong luminescence in the presence of light [9]. Further, the presence of cyanuric acid can cause aggregation of AuNPs during the reaction. AgNPs can also exhibit LSPR in the presence of proteins from *E. coli* and sodium ions. It is observed that the dispersity of AgNPs and the accuracy of colorimetric assays is significantly affected by the interaction with *E. coli* [9].

There are three mechanisms involved in the detection of analytes in label-based colorimetric techniques: These are: (1) non-cross-linking, (2) cross-linking, or (3) destabilization-induced aggregation [1]. The ligand-functionalized nanoparticles (f-NPs), in particular can trigger nanoparticle aggregation. The electrostatic repulsion phenomenon is inhibited from forming intermolecular bonds between ligands and analytes in the cross-linking mechanism as shown in (Fig. 1.2a). Aggregation on the other hand, can be observed when f-NPs are connected through the analytes depicted in (Fig. 1.2b), or an analyte can serve as attaching points for various f-NPs as shown in Fig. 1.2c. The shifts in LSPR can be observed due to the packing density increase of f-NPs regardless of their pathways.

LSPR can be compared to a highly-packed concerted crowd moving synchronously to the beat of the music. The collective oscillation of surface electrons of gold nanoparticles (AuNPs) or silver nanoparticles (AgNPs) will create a pattern of energy when interacted with lights that can be likened to a concert waves. This interaction could trigger color changes due to enhanced absorption of light. As a result, this phenomenon can be utilized as colorimetric sensor in various applications by observing the shift of their color changes.

Various applications can be identified utilizing AuNPs and AgNPs:

1. Rapid Diagnostic Kits: AuNPs and AgNPs have been utilized in heavy metal monitoring, detection for food safety pathogens, COVID-19 detection and rapid diagnostics in pregnancy tests.
2. Environmental Monitoring: Detection of toxic metals in water; AgNP sensor can exhibit rapid color change when water contamination is present.
3. Food Safety: AgNPs sensors are utilized for easy monitoring of contamination observed in color changes signaling the presence of toxic metals.
4. Healthcare Innovations: AgNPs are being used to detect cancer biomarkers for early diagnosis offering simple and cost-effective tests for urine and blood samples.
5. Educational and Portable Tools: AgNPs are utilized for healthcare diagnostics and remote environmental evaluations using smartphone apps that would enhance healthcare monitoring in remote area.

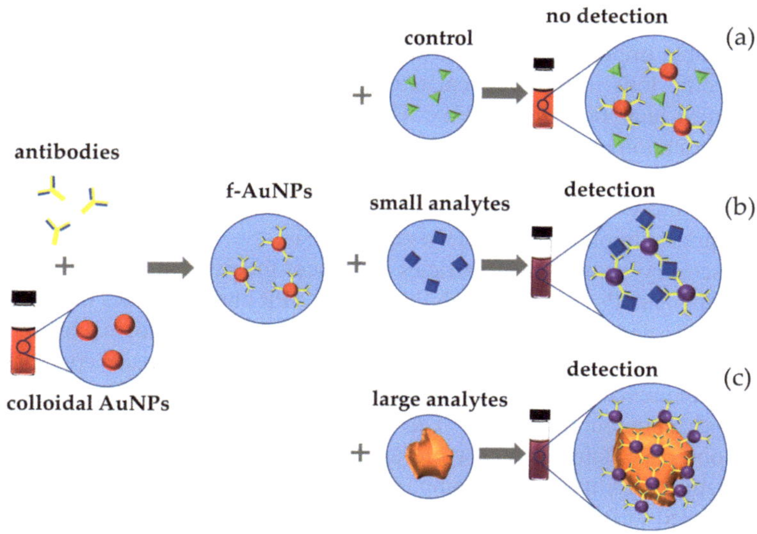

Fig. 1.2 Mechanism of detection for colorimetric immunosensor with antibody-functionalized nanoparticles: **a** No color change indicating that the analyte is not recognized by the nanoparticles, **b** nanoparticle aggregations are observed due to the presence of connected small analytes, **c** plasmon coupling is observed due to the binding of multiple nanoparticles with large analytes. Adapted from [1],Fig. 1.2. Licensed under CC BY 4.0 (https://creativecommons.org/licenses/by/4.0/)

1.2.1 Introduction to Colorimetric Sensing

Recently, significant advancement in analytical chemistry has been observed in colorimetric sensing applications through visual observation of color changes in the presence of analytes. Portable device technologies, in particular will always consider simplicity, accessibility and cost-effectiveness as compared to conventional analytical techniques. This is very important and remarkable to address various challenges in the field of food safety for public health protection and environmental pollution monitoring [6].

This technique demonstrates advancement in practical application and scientific innovation because, in actual real-world applications, it is important to equip researchers and practitioners with advanced tools to ensure accurate and reliable analyses. It is important to note that in handling complex chemical and biological samples, accuracy and simplicity are always considered. It is believed that colorimetric sensing not only pushes the limits of analytical chemistry but also symbolizes innovative imaginations that would address global needs and provide lasting and transformative solutions. These methods will effectively identify both the presence and concentration of various compounds, including heavy metals, fluoride, and hydrogen peroxide, to determine analyte levels in the sample [20].

The colorimetric sensing methods have been used over a long period of centuries [9]. Noticeable improvements since the nineteenth century have been observed to detect specific ions using chemical reagents. This method was the basis for the foundation of analytical colorimetry. This field undergoes various improvements and transformations ranging from qualitative evaluations to modern, precise, and quantitative techniques that are now being utilized in various pharmaceutical food analyses. In the twentieth century, photoelectric colorimeters were developed to facilitate quick, simple, and precise visual analysis.

The discovery of metal nanoparticles influences the modern field of colorimetric sensing using AuNPs as well as AgNPs having LSPR characteristics that can potentially enhance selectivity and accuracy of detection. This technological breakthrough offers a better alternative for equipment-free detection for on-site applications to assess environmental pollutant in soil, food safety monitoring, and detecting the presence of toxic compounds in water. The outstanding characteristics of AuNPs make them a good candidate for real-world applications due to their adjustable optical characteristics, stability, and biocompatibility. Recent advancements in lateral flow assays have led to faster and more accurate diagnostic tools for detecting COVID-19, alongside the development and fabrication of sensors capable of detecting heavy metals and organic environmental pollutants [21].

1.2.2 The Role of Metal Nanoparticles in Colorimetric Platforms

AuNPs and AgNPs play a vital role in the field of colorimetric sensing due to their unique optical characteristics that exhibit LSPR property. The LSPR phenomenon will trigger the observation of color changes. AuNPs, in particular, exhibit color changes from red to blue while AgNPs demonstrate color changes from violet to black depending on the state and amount of their respective metal aggregation [20]. The accuracy and selectivity of detection of these nanoparticles can be improved via the presence of aptamers, ligands, or polymer coatings functionalization, which will ultimately dictate the level of detection in a wide range of concentrations even up to the femtogram levels [11]. It is noteworthy to consider various aspects related to environmental concerns as well as understanding of the aggregation of these nanoparticles [21].

It is also very important to consider the environmental parameters such as temperature, pH, and ionic strength that will tend to destroy the stability of nanoparticles, leading to the formation of agglomerated nanoparticles that can comprise its stability in various solutions [18]. The issue of instability can cause problems in real-world complex biological or environmental applications. To address these problems, there is a need to introduce stabilizers often called "repeaters" or "probes". These stabilizers often include polyethylene glycol (PEG), surfactants such as citrate or polyvinylpyrrolidone, and specific ligands like disulfides or thiols. These stabilizing agents will possibly suppress aggregation and ensure the stability of the nanoparticles. As a result, the presence of these stabilizers could ensure accurate, reliable, and efficient results in various conditions.

The stability of nanoparticles can be enhanced with the use of stabilizing agents that will produce a good result when used in sensor in complex and dynamic environments. For example, to specifically detect heavy metals or biomolecules in different settings, it is important to introduce aptamer-functionalized gold nanoparticles to ensure stability and accuracy. Further, careful management of analyte-induced structural variations should also always be considered to ensure accurate and efficient reproducible colorimetric results.

1.3 Advantages of Metal Nanoparticle-Based Colorimetric Systems

Metal nanoparticles (MNP)-based colorimetric sensors have a wide range of applications and offer key advantages such as high accuracy and scalability. To enable the fabrication of these sensors, various synthesis techniques need to be employed. The chemical reduction method, specifically the Turkevich-Frens method, can be a good route to synthesis and can produce uniform AuNPs. Utilization of sodium citrate and gold ions is required for the reduction process. It involves the reduction and stabilization of ions to produce nanoparticles [3]. Seed-mediated growth method can be an alternative process that enables precise control of the size and morphology of nanoparticles. This method usually produces a diverse morphology of nanoparticles such as nanospheres, nanorods, and nanoflowers using specific analytes and optimized conditions [26].

Sustainable green synthesis techniques, on the other hand, are a good alternative for scalability in production utilizing biocompatible substances and plant extracts. This process could minimize environmental pollution and eventually reduce manufacturing costs [12]. In order to produce high-volume output, these scalable strategies can be explored. The synthesized nanoparticles using these strategies can be a good candidate for real-time monitoring of heavy metal pollutants in water, detection of pathogens for food safety applications, and the fabrication of rapid diagnostic kits for medical applications [16].

For specific example to fabricate sensors that are highly selective for specific pathogens, AuNPs functionalized with aptamers can be fabricated for this purpose. On the other hand, AgNPs are being utilized for environmental applications for the detection of toxins and heavy metals. It is important to note that the composition and functionalization of nanoparticles can be adjusted to achieve optimal results for specific purposes. There is wide array of applications both in environmental and biological aspects that would facilitate the prospects for commercial viabilities of the colorimetric systems [22].

1.3.1 Sensitivity, Selectivity, and Detection Limits

Sensitivity and accuracy are critical advantages of using metal nanoparticles in colorimetric systems, particularly for applications requiring ultra-low detection limits to identify target analytes. Utilization of these nanoparticles can detect minute environmental changes by exhibiting colorimetric shifts and altering their optical properties in order to measure an extremely low concentration of target analytes. Specific applications in food safety, environmental monitoring, and clinical diagnostics always require this capability. Furthermore, selectivity and accuracy of detection are very essential especially in the detection of analytes within complex mixtures [7, 30]. In colorimetric systems, the advancement in nanoparticle research has led to more advanced detection and even improved selectivity, accuracy, and reliability of analyses.

There are specific advantages in the testing process of utilizing metal nanoparticle-based sensing platforms in colorimetry. The change in color of the colorimetric sensor is most often analyzed using commercial scanners, digital cameras, UV–Vis spectrophotometers, and smartphone-based detectors. To validate the reliability of this sensing methodology, lower detection limits and selectivity to detect target analytes should always be considered. It is important to note that the mentioned colorimetric sensing techniques are rapid and user-friendly, with results that can be easily observed visually.

1.3.2 Practical Applications in Biosensing, Environmental Monitoring, and Food Safety

There are numerous applications of metal nanoparticle-based colorimetric systems. These include the detection of toxic molecules, heavy metals, pathogens, pharmaceuticals, and even biomarkers and urinary tract infections (UTIs) [15, 27]. Early detection of major diseases relies on the use of reliable and standardized protocols. An example of this detection and diagnosis of UTIs using metal oxide nanoparticles as shown in Fig. 1.3. It is believed that these methods are very important in environmental studies. Field monitoring often require portable detectors and rapid response in detection. As such, electroanalytical methods offer enduring advantages, including high selectivity, low cost, simplicity, and rapid results, making them a strong alternative to conventional analytical techniques. Environmental safety and water management must require simple and fast detection techniques.

The ability to deliver readily available, real-time results is critical for making swift managerial decisions. The versatility of fabricated metal nanoparticles in biosensing provides significant advantages, including a large surface area that enables rapid detection and responsiveness to a wide range of environmental analytes [5, 19]. Growing global healthcare demands for biomarker diagnostics are driving the development of colorimetric biosensing systems to meet these needs. This advancement

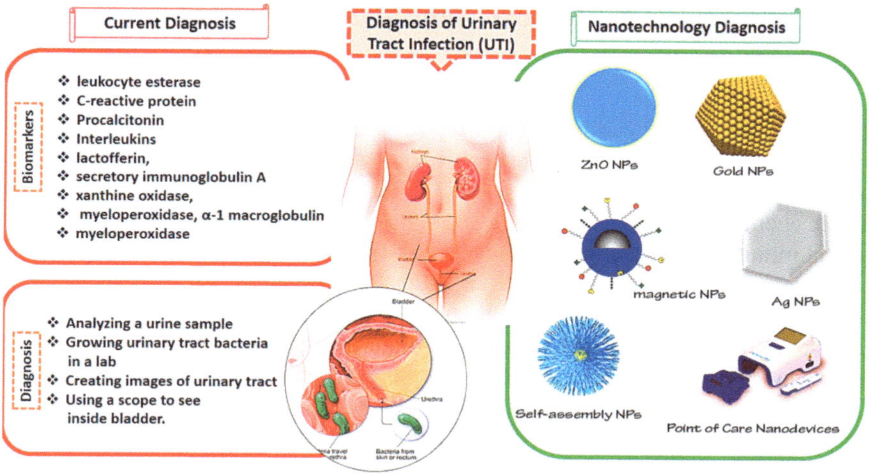

Fig. 1.3 Recent methods and metal oxide approaches in diagnosing urinary tract infections (UTIs). Adapted from [17], Fig. 1.3. Licensed under CC BY 4.0 (https://creativecommons.org/licenses/by/4.0/)

has significantly impacted rapid contaminant detection in food safety, benefiting both consumers and producers through the availability of affordable, point-of-use detectors capable of analyzing optically transparent solids and dispersions. Point-of-care testing technologies are also being explored for future innovation.

1.4 Objectives and Scope of the Book

This book can be used as a comprehensive reference for students and professionals in the fields of engineering, biomedical science, and environmental science about colorimetric sensing utilizing metal nanoparticles. A foundational understanding that bridges theoretical concepts with real-world applications are presented, emphasizing collaboration between scientists and engineers to address existing challenges in the field. Colorimetric sensors are highly preferred due to their sensitivity, ease of use, and rapid response. Researchers are increasingly focusing on controlling the size of metal nanoparticles, as this is critical for plasmon-induced color changes essential to advanced detection methods.

1.4.1 Purpose of the Book in Consolidating Current Advancements

Recently, researchers worldwide have reported significant progress in colorimetric sensing by leveraging the optical properties of carefully selected metal nanoparticles in the fabrication of colorimetric sensors. This book presents the development of simple chemical to advanced colorimetric sensors as applied to biosensing, environmental monitoring, and clinical diagnoses. The significant advancements in colorimetric sensors are presented in three key steps: (a) developing a strategy to optimize the complex properties of metal nanoparticles, (b) recognizing that several real-world analyses remain unaddressed by current sensor technologies or smart devices, and (c) exploring various methods to fabricate colorimetric sensors for diverse applications. This book highlights the sensing capabilities of metal nanoparticles for potential use in advanced sensors and nanoplatforms in real-world applications.

1.4.2 Outline of the Book Structure and Organization

The topic selection included in this book has been carefully chosen to address the important aspects in the development and actual implementation of metal nanoparticles utilized as colorimetric sensors. The book starts with a chapter on introducing the concepts of a colorimetric sensor utilizing metal nanoparticles, while Chap. 2 explores the specifics of utilizing gold nanoparticles as a colorimetric sensor. Chapter 3, on the other hand, introduces the role of copper nanoparticles fabricated as a colorimetric sensor. Further, Chap. 4 introduces the audience to the topic on enhanced colorimetric sensing detection using iron nanoparticles. This book continues with Chap. 5, that focuses on silver, platinum, and palladium nanoparticles utilized as colorimetric sensors. Chapter 6 introduces the audience to the topic on hybrid and composite metal nanoparticles as applied in colorimetric systems. Additionally, Chap. 7 emphasizes the real-world application in environmental analysis, medical diagnostics, and biosensing. Chapter 8 delves into the challenges, innovations, and recommendations in the use of MNPs for colorimetric sensing. Lastly, Chap. 9 presents the concluding remarks that include the suggested future directions and insights of MNPs for colorimetric detection.

References

1. Acunzo, A., Scardapane, E., De Luca, M., Marra, D., Velotta, R., Minopoli, A.: Plasmonic nanomaterials for colorimetric biosensing: a review. Chemosensors **10**(4), 136 (2022). https://doi.org/10.3390/chemosensors10040136
2. Aghababaie, M., Foroushani, E.S., Changani, Z., Gounani, Z., Mobarakeh, M.S., Hadady, H., Razmjou, A.: Recent advances in the development of enzymatic paper-based microfluidic

biosensors. Biosens. Bioelectron.. Bioelectron. **226**, 115131 (2023). https://doi.org/10.1016/j.bios.2023.115131
3. Alberti, G., Zanoni, C., Magnaghi, L.R., Biesuz, R.: Gold and silver nanoparticle-based colorimetric sensors: new trends and applications. Chemosensors **9**, 305 (2021). https://doi.org/10.3390/chemosensors9110305
4. Chen, Z., Zhang, Z., Qi, J., You, J., Ma, J., Chen, L.: Colorimetric detection of heavy metal ions with various chromogenic materials: strategies and applications. J. Hazard. Mater. **441**, 129889 (2023). https://doi.org/10.1016/j.jhazmat.2022.129889
5. Fritea, L., Banica, F., Costea, T.O., Moldovan, L., Dobjanschi, L., Muresan, M., Cavalu, S.: Metal nanoparticles and carbon-based nanomaterials for improved performances of electrochemical (Bio) sensors with biomedical applications. Materials **14**(21), 6319 (2021). https://doi.org/10.3390/ma14216319
6. Garrido-Maestu, A., Prado, M.: Naked-eye detection strategies coupled with isothermal nucleic acid amplification techniques for the detection of human pathogens. Compr. Rev. Food Sci. Food Saf. **21**(2), 1913–1939 (2022). https://doi.org/10.1111/1541-4337.12902
7. Gordón, J., Arruza, L., Ibáñez, M.D., Moreno-Guzmán, M., López, M.Á., Escarpa, A.: On the move-sensitive fluorescent aptassay on board catalytic micromotors for the determination of interleukin-6 in ultra-low serum volumes for neonatal sepsis diagnostics. ACS Sens. **7**(10), 3144–3152 (2022). https://doi.org/10.1021/acssensors.2c01635
8. Han, F., Wang, T., Liu, G., Liu, H., Xie, X., Wei, Z., Xu, F.: Materials with tunable optical properties for wearable epidermal sensing in health monitoring. Adv. Mater. **34**(26), 2109055 (2022). https://doi.org/10.1002/adma.202109055
9. Hua, Z., Yu, T., Liu, D., Xianyu, Y.: Recent advances in gold nanoparticles-based biosensors for food safety detection. Biosens. Bioelectron.. Bioelectron. (2021). https://doi.org/10.1016/j.bios.2021.113076
10. Hyder, A., Buledi, J.A., Nawaz, M., Rajpar, D.B., Orooji, Y.: Identification of heavy metal ions from aqueous environment through gold, silver and copper nanoparticles: an excellent colorimetric approach. Environ. Res. **205**, 112475 (2022). https://doi.org/10.1016/j.envres.2021.112475
11. Khani, H., Abbasi, S., Yaraki, M.T., Tan, Y.N.: A naked-eye colorimetric assay for detection of Hg^{2+} ions in real water samples based on gold nanoparticles-catalyzed clock reaction. J. Mol. Liq. (2022). https://doi.org/10.1016/j.molliq.2021.118243
12. Kumar, S., Singh, R.: Recent optical sensing technologies for the detection of various biomolecules. Opt. Laser Technol. (2021). https://doi.org/10.1016/j.optlastec.2020.106620
13. Liu, B., Zhuang, J., Wei, G.: Recent advances in the design of colorimetric sensors for environmental monitoring. Environ. Sci. Nano (2020). https://doi.org/10.1039/D0EN00449A
14. Luo, X., Zaitoon, A., Lim, L.T.: A review on colorimetric indicators for monitoring product freshness in intelligent food packaging: indicator dyes, preparation methods, and applications. Compr. Rev. Food Sci. Food Saf. **21**(3), 2489–2519 (2022). https://doi.org/10.1111/1541-4337.12942
15. Nilghaz, A., Mousavi, S.M., Tian, J., Cao, R., Guijt, R.M., Wang, X.: Noble-metal nanoparticle-based colorimetric diagnostic assays for point-of-need applications. ACS Appl. Nano Mater. **4**(12), 12808–12824 (2021). https://doi.org/10.1021/acsanm.1c01545
16. Putri, L.A., Prabowo, Y.D., Dewi, D.M.M., Mumtazah, Z., Adila, F.P., Fadillah, G., Wasisto, H.S.: Review of noble metal nanoparticle-based colorimetric sensors for food safety monitoring. ACS Appl. Nano Mater. **7**(17), 19821–19853 (2024). https://doi.org/10.1021/acsanm.4c04327
17. Qindeel, M., Barani, M., Rahdar, A., Arshad, R., Cucchiarini, M.: Nanomaterials for the diagnosis and treatment of urinary tract infections. Nanomaterials **11**(2), 546 (2021). https://doi.org/10.3390/nano11020546
18. Rajkumar, G., Sundar, R.: Sonochemical-assisted eco-friendly synthesis of silver nanoparticles (AgNPs) using avocado seed extract: Naked-eye selective colorimetric recognition of Hg^{2+} ions in aqueous medium. J. Mol. Liq. (2022). https://doi.org/10.1016/j.molliq.2022.120638

19. Ramesh, M., Janani, R., Deepa, C., Rajeshkumar, L.: Nanotechnology-enabled biosensors: a review of fundamentals, design principles, materials, and applications. Biosensors (2022). https://doi.org/10.3390/bios13010040
20. Saadati, A., Farshchi, F., Hasanzadeh, M., Liu, Y., Seidi, F.: Colorimetric and naked-eye detection of arsenic (iii) using a paper-based microfluidic device decorated with silver nanoparticles. RSC Adv. (2022). https://doi.org/10.1039/d2ra02820d
21. Sadiq, S., Khan, S., Khan, I., Khan, A., Humayun, M., Wu, P., Usman, M., Khan, A., Alanazi, A.F., & Bououdina, M.: A critical review on metal-organic frameworks (MOFs) based nanomaterials for biomedical applications: Designing recent trends challenges and prospects. Heliyon **10**(2), e25521 (2024). https://doi.org/10.1016/j.heliyon.2024.e25521
22. Scroccarello, A., Della Pelle, F., Del Carlo, M., Compagnone, D.: Optical plasmonic sensing based on nanomaterials integrated in solid supports. A critical review. Anal. Chim. ActaChim. Acta **1237**, 340594 (2023). https://doi.org/10.1016/j.aca.2022.340594
23. Semin, S., Li, X., Duan, Y., Rasing, T.: Nonlinear optical properties and applications of fluorenone molecular materials. Adv. Opt. Mater. (2021). https://doi.org/10.1002/adom.202100327
24. Singh, H., Bamrah, A., Bhardwaj, S.K., Deep, A., Khatri, M., Brown, R.J., Kim, K.H.: Recent advances in the application of noble metal nanoparticles in colorimetric sensors for lead ions. Environ. Sci. Nano **8**(4), 863–889 (2021). https://doi.org/10.1039/D0EN00963F
25. Umapathi, R., Sonwal, S., Lee, M.J., Rani, G.M., Lee, E.S., Jeon, T.J., Huh, Y.S.: Colorimetric based on-site sensing strategies for the rapid detection of pesticides in agricultural foods: new horizons, perspectives, and challenges. Coord. Chem. Rev.. Chem. Rev. **446**, 214061 (2021). https://doi.org/10.1016/j.ccr.2021.214061
26. Wang, H., Wu, T., Li, M., Tao, Y.: Recent advances in nanomaterials for colorimetric cancer detection. J. Mater. Chem. B (2021). https://doi.org/10.1039/D0TB02163F
27. Yang, J., Wang, X., Sun, Y., Chen, B., Hu, F., Guo, C., Yang, T.: Recent advances in colorimetric sensors based on gold nanoparticles for pathogen detection. Biosensors **2023**(13), 29 (2022). https://doi.org/10.3390/bios13010029
28. Yao, Z., Coatsworth, P., Shi, X., Zhi, J., Hu, L., Yan, R., Yu, H.D.: Based sensors for diagnostics, human activity monitoring, food safety and environmental detection. Sens. Diagn. **1**(3), 312–342 (2022). https://doi.org/10.1039/d2sd00017b
29. Yu, L., Song, Z., Peng, J., Yang, M., Zhi, H., He, H.: Progress of gold nanomaterials for colorimetric sensing based on different strategies. TrAC, Trends Anal. Chem. **127**, 115880 (2020). https://doi.org/10.1016/j.trac.2020.115880
30. Zamanian, J., Khoshbin, Z., Abnous, K., Taghdisi, S.M., Hosseinzadeh, H., Danesh, N.M.: Current progress in aptamer-based sensing tools for ultra-low level monitoring of Alzheimer's disease biomarkers. Biosens. Bioelectron. **197**, 113789 (2022). https://doi.org/10.1016/j.bios.2021.113789

Chapter 2
Gold Nanoparticle-Based Colorimetric Detection Systems

Abstract Colorimetric sensors based on gold nanoparticles (AuNP) have attracted great attention in analytical chemistry due to their extraordinary sensitivity and versatility. This chapter explores quasi-two-dimensional nanoparticles and metallic nanoparticles, highlighting how their characteristic features such as size, shape, and low extinction coefficients contribute to their localized surface plasmon resonance (LSPR) optical properties in response to light. These unique properties allow AuNPs to sense trace amounts of analytes, where colorimetric changes act as a visible signal. However, several challenges such as nanoparticle agglomeration, environmental sensitivity, and limited quantitative detection still exist. To improve stability, selectivity, and accuracy of these AuNPs, surface functionalization with stabilizing agents such as polyethylene glycol (PEG) polymers and thiol groups has been suggested. This chapter goes through the different reaction mechanisms of AuNP-based colorimetric sensors, with applications for environmental monitoring, medical diagnostics, and food safety. Moreover, this chapter also aims to discuss sensor efficiency improvement strategies, including both aggregation-based and non-aggregation-based detection approaches. Despite these current limitations, the use of AuNP-based sensors is poised to promote real-time, on-site detection of targeted analytes in various applications.

Colorimetric sensors have been intensively studied and have attracted researchers around the globe owing to their promise and potential of lowering production costs and their simplicity of fabrication methods [9, 41]. Size distribution, morphology, and high extinction coefficient are among the properties of gold nanoparticles (AuNPs) being considered in the fabrication of colorimetric sensors because they influence and trigger color changes. AuNPs will detect analytes via weak interactions in hydrogen bonding, covalent bonds, and electrostatic interactions [5, 8]. The colorimetric signal can be detected as a result of the interaction between AuNPs and the analytes, which produces a reaction product – an essential property in designing colorimetric detection systems [23, 37].

Many important and practical applications of AuNP-based colorimetric detection stem from its high sensitivity and selectivity. Limitations of utilizing AuNPs

A. C. Alguno et al., *Metal Nanoparticles-Based Nanoplatforms for Colorimetric Sensing*, Engineering Materials, https://doi.org/10.1007/978-981-96-6771-0_2

as a colorimetric sensor are as follows: (a) complications on its operation; (b) presence of AuNP aggregation during their applications in urine, blood, or seawater; and (c) imprecision in quantitative detection. Improvement of these sensors is needed especially when it is utilized for environmental monitoring, disease diagnosis, and food safety applications. Usually, a higher concentration detection limit is achieved in colorimetric detection compared to other methods. It is, therefore, imperative and important to consider the appropriate choice of the detection systems to meet the expected detection limit which is affected by the hardware or colorimetric sensors in the systems.

2.1 Properties of Gold Nanoparticles for Sensing

The size of AuNPs ranges from 1 to 100 nm, with their electronic, chemical, and optical properties being size-dependent. Likewise, the localized surface plasmon resonance (LSPR) is also dependent on these properties and is environmentally sensitive. It is for this reason why AuNPs are being utilized in medical diagnostics, catalysis, environmental sensors, and other sensing applications. There are various types of AuNPs as shown in Fig. 2.1.

AuNPs can be a good and useful nanomaterials for colorimetric sensor of analytes in various applications due to their capability to improve optical signals. It has been reported that due to AuNPs, optical signals extinction coefficient with a typical value

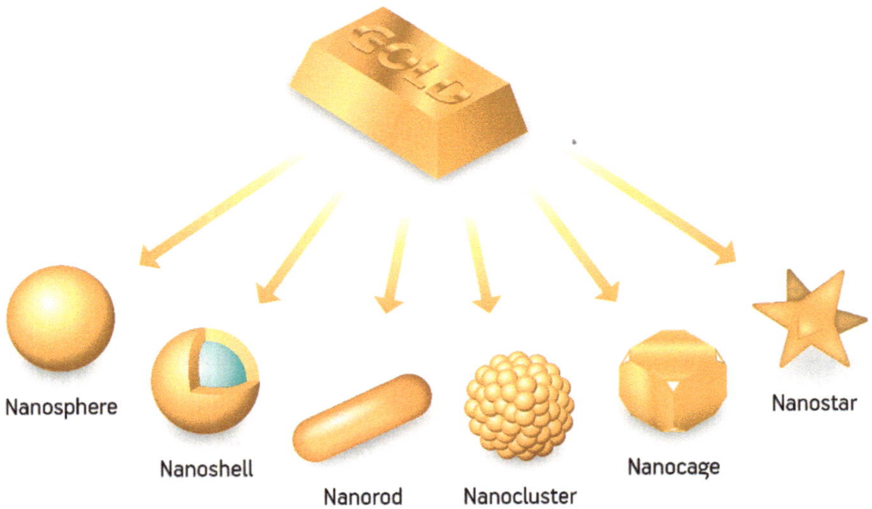

Fig. 2.1 Various types of gold nanoparticle structures and their corresponding morphologies. Adapted from [14], Fig. 2.1. Licensed under CC BY 4.0 (https://creativecommons.org/licenses/by/4.0/)

between (6–9) $\times$ 10^8 M^{-1} cm^{-1} can be achieved providing a quite better sensitivity than organic dyes in the detection of analytes in a wide array of applications. Nevertheless, several challenges still persist especially at higher concentrations of the analyte being detected. One of the challenges is the aggregation of AuNPs that eventually affects their sensitivity and accuracy which will contribute to the problem of analyte detection reproducibility. This issue has been reported in heavy metal detection and glucose biosensing, where high accuracy and sensitivity are essential for measuring quantitative concentrations [32].

Furthermore, AuNPs are very sensitive to various environmental factors including temperature, pH, and ionic strength. The changes in environmental conditions could also influence the nanoparticles' dielectric property as well as their corresponding surface plasmon resonance (SPR) stability. This may lead to false positive results in biosensing applications, as high ionic strength can cause the aggregation of AuNPs [1]. The aforementioned problems can be mitigated by modifying the surface with the introduction of stabilizing agents.

To improve the stability of AuNPs, polyethylene glycol (PEG) polymer coatings have been used to improve its performance in complex biological environments. Moreover, it has been suggested that functionalizing ligands, particularly with thiol or disulfide groups, can enhance selectivity and accuracy by minimizing non-specific interactions [6]. These methods will provide high sensitivity for specific analytes and broaden the application of AuNPs in biomedical and food safety fields, enabling the development of reproducible and reliable detection processes.

2.1.1 High Extinction Coefficient and Tunable Optical Properties

High extinction coefficient. A high extinction coefficient is one of the key properties of metal nanoparticles (MNPs), making them highly effective in the visible and near-infrared regions. It represents the combined effect of scattering and absorption cross-sections at specific wavelengths [18, 21]. The combined scattering and absorption properties of MNPs significantly influence their colorimetric performance, which is critical for real-world applications such as photothermal therapy and colorimetric sensing. When light interacts with these nanoparticles, it generates a SPR band, which subsequently gives rise to surface plasmons. Additionally, AuNPs are believed to offer exceptionally sensitive detection exceeding the extinction coefficients of organic dyes and quantum dots by more than an order of magnitude.

Tunable optical properties. The tunable optical properties of MNPs are inherent to their structure, which consists of ultra-small metallic particles. Factors such as shape, size, the dielectric constant of the surrounding medium, and the chemical composition of the nanoparticles all contribute to their distinct optical behavior [22, 25]. The surface plasmon resonance peak is usually observed in the near-infrared regions with specific peaks found around tens to hundreds of nanometers [26, 38]. Enhanced accuracy and sensitivity largely depend on the aforementioned factors, which also contribute to the overall performance of colorimetric sensors. For instance, changes in absorption color are influenced by interactions between nanoparticles, which alter plasmonic coupling. However, tuning the resonance peak can be challenging, as it is also affected by the properties of the surrounding medium.

2.2 Mechanisms of Gold Nanoparticle-Based Colorimetric Reactions

The effective operation of AuNP-based colorimetric sensors relies on localized surface plasmon resonance (LSPR), which is influenced by nanoparticle aggregation and the choice of substrates [2, 8, 37]. These characteristics and principles are key considerations in the fabrication of AuNP-based colorimetric sensors. Studies have shown that traditional chemical dyes offer lower stability compared to AuNPs. The optical properties of AuNPs are influenced by factors such as shape, size, aggregation state, and particle composition. Additionally, the dielectric constant of the surrounding medium plays a significant role in determining their optical behavior. All these parameters can be precisely tailored and fine-tuned for practical applications, including the detection of hazardous materials[24].

AuNPs are widely available commercially, largely due to their regulatory approval in many countries, which enhances their marketability. These nanomaterials are readily accessible, as most regulatory bodies have cleared their use. Typically, the smallest AuNPs are about 10 nm in size and exhibit a spherical shape, with a maximum plasmon absorption band in the range of 500–510 nm. However, as the nanoparticle size increases, the plasmon absorption bands shift due to an increase in conduction band electron density, which alters their optical properties [13, 22].

The overall properties of AuNPs in fabricated colorimetric detection systems are influenced by several factors, including particle shape and composition, size, the dielectric constant of the surrounding medium, and surface functional groups [30, 34] . The sensitivity and accuracy of fabricated AuNP colorimetric sensors are highly dependent on their size. Research has shown that larger AuNPs tend to exhibit a blueshift in the plasmon peak, which results in a less sensitive sensor.

Moreover, to effectively capture Hg^{2+}, there is a need to control the AuNP shape and their corresponding composition. Citrate-capped AuNPs offer a distinct advantage for this application [3, 4, 17]. In citrate-capped AuNPs, the thiol groups would produce a larger particle. Optimization using aptamers would provide maximum LSPR wavelength sensitivity. The dependence on the surface charge of the materials and the reacting analyte is strong and affects the transmission spectrum and LSPR wavelength shift [16, 42]. The pH and ion strength of the environmental condition could also affect the properties of the synthesized AuNPs. It has been reported that the application of AuNPs for detecting amino acids in serum solutions showed limited efficiency, primarily due to the complexity involved in the isolation process [19, 27].

The schematic diagram of the mechanisms of AuNP-based colorimetric sensors utilized in various applications is shown in Fig. 2.2. Figure 2.2 illustrates the mechanism of a multifunctional Hg^{2+} –sensitive sensor, which employs a specific aptamer to selectively bind the target during the AuNP-mediated growth process. The aptamer will be detached from AuNP when Hg^{2+} ions comes in contact with the aptamers. The quantitative detection of sulfadimethoxine (SDM), on the other hand, is shown in Fig. 2.2b, which illustrates the AuNP aggregation-induced sensor applied in smartphones. The assay used in this application prevents AuNP aggregation in the presence of the aptamer, as the binding of SDM to the aptamer occurs on the AuNP surface.The visual method for detecting live *Salmonella typhimurium* utilizing Au nanostars (AuNSPs) is illustrated in Fig. 2.2c. Aptamer 2 provides a complex sandwich network with Aptamer 1 and the target, while Aptamer 1 easily binds to the target. Lastly, the instantaneous detection of tetracycline and chloramphenicol using AuNP aggregation coordinated with single-stranded DNA fragments in the fabricated colorimetric sensor is shown in Fig. 2.2d.

2.2.1 Aggregation and Non-aggregation-Based Detection Strategies

To improve the performance and detection properties of colorimetric sensors, a combination of non-aggregation- and aggregation-based techniques have been employed [36]. These techniques have their advantages and disadvantages. It is expected that aggregation detection systems exhibit a higher sensitivity. They are less likely influenced by the unwanted protein adsorption that generally covers the critical binding sequences. Usually, non-aggregation-based sensors are inferior in terms of performance as compared to aggregation-based sensors when applied to unknown sample analyses except in multiplexing conditions [35]. In various detection systems, non-aggregation-based techniques usually allow extensive multiplexing that immediately allows the scanning of multiple surfaces [23]. On the other hand, aggregation-based sensors may lack the flexibility to perform reliably under varying conditions. In contrast, non-aggregation-based detection is often preferred in bioimaging applications, as it offers improved specificity and sensitivity, making it more effective for real-world applications.

2.2.2 Modifications and Functionalization for Enhanced Performance

To achieve an enhanced selectivity and sensitivity of the fabricated colorimetric sensor, functionalization as well as chemical and physical alterations of AuNPs should be performed [8]. The AuNP surface should be functionalized using functional

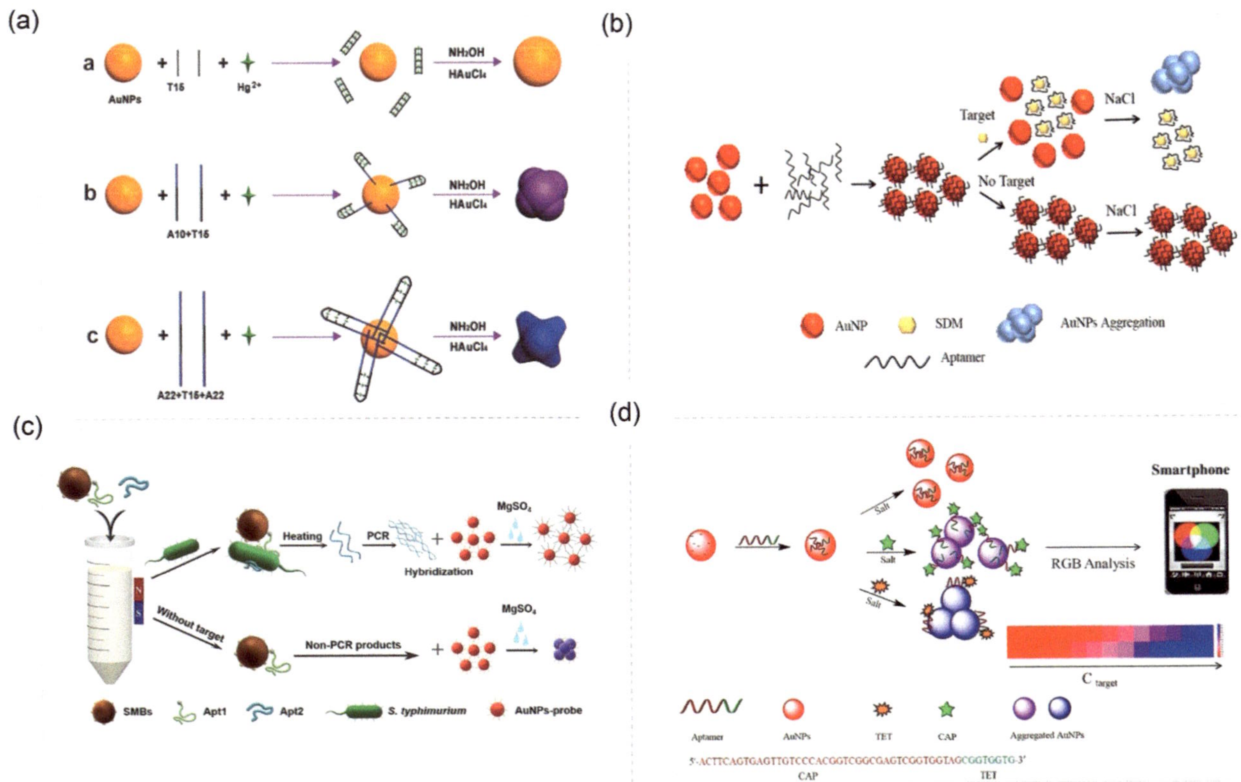

Fig. 2.2 **a** Colorimetric sensors that are highly sensitive to Hg^{2+} ions using AuNP with aptamer binding. **b** Controlling the non-aggregation of AuNP using aptamer to detect sulfadimethoxine (SDM) via a smartphone-based detection system. **c** Complex aptamer sandwich network with AuNSPs visually detects *Salmonella typhimurium*. **d** AuNP aggregation has been regulated by DNA to perform multi-target such as tetracycline chloramphenicol. Adapted from [11], Fig. 2.2.

groups and coated with oxides, surfactants, polymers, and graphene shells. The aim of these strategies is to achieve the dynamic range of detection and enhance the selectivity and sensitivity of the fabricated sensors. It is also important to consider the response time, detection limit values, and improvement of its stability in the functionalization process.

The functionalization and modification of AuNPs will significantly affect their surface chemistry and size distribution. The available number of binding sites of AuNPs will determine the response of detection for inorganic anions, metal ions, and volatile organic compounds. A wide range of applications of sensor technology are expected due to the presence of these new materials. Furthermore, development of highly sensitive AuNPs biosensors has been developed using the concepts of bioconjugation chemistry [33]. The sensitivity of these sensors even reached a very low limit of detection which is very sensitive up to the attomolar and femtomolar levels of molar concentrations. Recent efforts are focused on the reduction of response time and detection capability [5]. Moreover, the use of PEG-modified biotin for co-functionalizing AuNPs enables the development of an ultra-selective colorimetric sensor for detecting biotinylated DNA [35]. This method will affect and influence the SPR response that will allow visual detection of the color change. This is important for understanding binding interactions in DNA biosensing applications, where high accuracy and sensitivity are essential.

2.3 Applications of Gold Nanoparticle-Based Sensors

MNPs have various applications in cell tracking, imaging, detection, and drug delivery as reported in the literature [12]. These MNPs can possibly be utilized to address future applications in medical use due to their high biocompatibility and unique properties, especially in the application of personalized medicine. Among these MNPs, AuNPs are highly preferred because of their sensitivity and versatility when applied in real-world applications due to their visible changes when they bind to the target analyte [29]. Applications like home and field diagnostics are particularly useful. Due to this development, several AuNP-based colorimetric sensors have been developed. This AuNP-based sensor exhibits a better performance as compared to conventional assays such as microRNA, and cytokine assays, in terms of reagent usage, specificity, accuracy, and ease of use [20]. AuNPs play a significant role in enhancing phototherapy, making them highly effective for theranostic and localized imaging applications. Their low-cost and simplicity in detection are key advantages for developing point-of-care testing systems, especially suited for low-resource environments.

2.3.1 Detection of Heavy Metals and Toxins

Among all environmental contaminants, heavy metals, metalloids, and toxins are among the most hazardous due to their high mobility, bioaccumulation potential, toxicity, and persistence in the environment. Lead and mercury compounds, including their organoderivatives, pose significant ecological risks [39]. The metals mentioned above are readily transported and deposited in water bodies from both natural and anthropogenic sources. Due to their high levels of toxicity – particularly their neurotoxic effects—they can bioaccumulate, posing serious health risks to fetuses, children, and adults. Given their significant environmental impact, various detection methods have been developed, with particular emphasis on innovative colorimetric nanosensors using noble metal nanoparticles. These sensors are highly effective for both qualitative and quantitative detection of heavy metals at trace levels [3, 4]. A study is currently underway to investigate how nanoparticle size and interparticle distance influence sensor performance, particularly in terms of visible color changes during interactions with analytes. These color changes serve as visual indicators of the presence of toxic substances in the environment and can be subsequently recorded and analyzed for detection and monitoring purposes [3].

2.3.2 Medical Diagnostics: Cancer Markers, Proteins, and Nucleic Acids

MNPs play a significant role in the specific detection of clinically significant analytes [15]. Bioassays for cancer marker determination use gold and silver nanoparticles, functionalized with DNA markers or antibodies. Signaling methods enhance detection limits, with nanoparticle color serving as an indicator of DNA strand positioning among billions of probes. To detect low-concentration targets, aptamer-labeled nanoparticles are employed particularly in protein detection and clinical diagnostics. The hybridization of MNPs with DNA or RNA is widely used for identifying specific nucleic acid sequences. Sensitivity can be further improved by integrating oligonucleotide-functionalized nanoparticles to target a variety of nucleic acids. Additionally, plasmon resonance imaging provides quantitative data by analyzing the affinity interactions of nanoparticle populations within a defined structure [31]. This method captures important light absorption properties without the need for nanoparticle labeling in the medium and operates rapidly, allowing for the distinction of unique absorption characteristics. It holds significant value in scientific research.

2.3.3 *Environmental Monitoring: Pesticide and Pollutant Detection*

The environment and natural resources are increasingly threatened by a wide range of harmful organic and inorganic pollutants, posing serious risks to both ecological systems and human health. This growing concern emphasizes the urgent need for accurate analytical tools to efficiently detect hazardous analytes. Among the most precise methods available is the use of LSPR-based colorimetric detection systems, which can identify analyte concentrations by monitoring color changes down to trace levels. This emerging technology shows strong potential for environmental monitoring, particularly in detecting pesticide residues in soil and water. The improper and frequent use of toxic pesticides has not only impacted health and ecosystems but has also contributed to the rise of pesticide-resistant pests—a major challenge for sustainable agriculture in the face of climate change. In this context, nanotechnology-based LSPR sensors offer a promising solution for reliable and sensitive detection of pesticide residues [40]. These sensors offer significant advantages due to their speed, low cost, and suitability for both on-site and off-site analysis. However, current literature primarily focuses on the detection of pesticide molecules without adequately correlating experimental results with theoretical models. This discussion aims to bridge that gap by exploring the theoretical applicability of LSPR probes, while also highlighting the need for future studies to quantitatively assess the concentrations of analytes involved. Such efforts would enhance the accuracy, credibility, and overall efficiency of LSPR-based detection systems.

2.4 Challenges and Prospects

MNP-based nanoplatforms have been proposed as colorimetric sensing tools, but there are several issues that prevent their practical application. The stability, reproducibility, and sensitivity of these nanoparticles must be carefully balanced and considered across the platform [7]. The stability of these nanoparticles is quite unpredictable and is influenced by environmental factors, which complicates the problem of long-term reproducibility. The importance of understanding stochastic perturbations that impact experimental design, test kits, and measurements is vital but remains challenging, which results in bottlenecks to sensing performance. Despite colorimetric sensing having a high utility potential, there are still some issues that should be addressed. Therefore, there is a need for advanced technologies that can improve the sensitivity in strategic areas. Based on current limitations, certain scenarios highlight the need for selective, disruption-resistant performance, which may be achieved through either empirical methods or advanced detection solutions [10]. As inquiry and intellectual collaboration gather pace, colorimetric sensing is expected to develop, with accurate information on MNP properties emerging, thereby demanding a thorough investigation of all potential areas.

2.4.1 Overcoming Limitations in Stability, Reproducibility, and Sensitivity

The optical properties of AuNPs such as LSPR will be utilized to enhance the detection of colorimetric sensors. However, the wider application of AuNPs fabricated as colorimetric sensors is limited due to their poor sensitivity, reproducibility, and low stability [28]. Environmental conditions consistently impact the optical stability of AuNPs. This challenge can be mitigated by selecting materials that are minimally affected by such external factors. Additionally, structural optimization during synthesis is important to counterbalance the negative effects of environmental influences and ensure reliable sensor performance.

Sustainability, toxicity, and production cost are important factors that need to be considered for mass production of the sensor. Given the known biocompatibility and nontoxicity of AuNPs, they are considered sustainable and environmentally friendly, especially when synthesized using non-toxic methods. AuNPs often exhibit superior sensing capabilities compared to other nanoparticles. While production costs remain a challenge, advancements in green chemistry hold promise for enabling scalable and cost-effective manufacturing processes.

References

1. Abdelhamied, M.M., Atta, A., Abdelreheem, A.M., Farag, A.T.M., El Sherbiny, M.A.: Oxygen ion induced variations in the structural and linear/nonlinear optical properties of the PVA/PANI/Ag nanocomposite film. Inorg. Chem. Commun.. Chem. Commun. **133**, 108926 (2021). https://doi.org/10.1016/j.inoche.2021.108926
2. Alberti, G., Zanoni, C., Magnaghi, L.R., Biesuz, R.: Gold and silver nanoparticle-based colorimetric sensors: New trends and applications. Chemosens. **9**(11), 305 (2021). https://doi.org/10.3390/chemosensors9110305
3. Ali, S., Chen, X., Shi, W., Huang, G., Yuan, L.M., Meng, L., Chen, X.: Recent advances in silver and gold nanoparticles-based colorimetric sensors for heavy metal ions detection: a review. Crit. Rev. Anal. Chem. **53**(3), 718–750 (2023). https://doi.org/10.1080/10408347.2021.1973886
4. Ali, Z., Ullah, R., Tuzen, M., Ullah, S., Rahim, A., Saleh, T.A.: Colorimetric sensing of heavy metals on metal doped metal oxide nanocomposites: a review. Trends Environ. Anal. Chem. **37**, e00187 (2023). https://doi.org/10.1016/j.teac.2022.e00187
5. Anh, N.H., Doan, M.Q., Dinh, N.X., Huy, T.Q., Tri, D.Q., Le, A.T.: Gold nanoparticles-based SERS nanosensor for thiram and chloramphenicol monitoring in food samples: Insight into effects of analyte molecular structure on their sensing performance and signal enhancement. Appl. Surf. Sci. **584**, 152555 (2022). https://doi.org/10.2139/ssrn.3986012
6. Arrico, L., Di Bari, L., Zinna, F.: Quantifying the overall efficiency of circularly polarized emitters. Chem.–A Eur. J. (2021). https://doi.org/10.1002/chem.202002791
7. Ateia, M., Wei, H., Andreescu, S.: Sensors for emerging water contaminants: overcoming roadblocks to innovation. Environ. Sci. Technol. **58**(6), 2636–2651 (2024). https://doi.org/10.1021/acs.est.3c09889
8. Chatterjee, S., Lou, X.Y., Liang, F., Yang, Y.W.: Surface-functionalized gold and silver nanoparticles for colorimetric and fluorescent sensing of metal ions and biomolecules. Coord. Chem. Rev.. Chem. Rev. (2022). https://doi.org/10.1016/j.ccr.2022.214461

9. Chen, Z., Zhang, Z., Qi, J., You, J., Ma, J., Chen, L.: Colorimetric detection of heavy metal ions with various chromogenic materials: strategies and applications. J. Hazard. Mater. **441**, 129889 (2023). https://doi.org/10.1016/j.jhazmat.2022.129889
10. Clauson, K.A., Crouch, R.D., Breeden, E.A., Salata, N.: Blockchain in pharmaceutical research and the pharmaceutical value chain. In: Blockchain in Life Sciences, pp. 25–52. Springer Nature Singapore, Singapore (2022). https://doi.org/10.1007/978-981-19-2976-2_2
11. Cui, Y., Zhao, J., Li, H.: Chromogenic mechanisms of colorimetric sensors based on gold nanoparticles. Biosensors **13**(8), 801 (2023). https://doi.org/10.3390/bios13080801
12. Desai, N., Momin, M., Khan, T., Gharat, S., Ningthoujam, R.S., Omri, A.: Metallic nanoparticles as drug delivery system for the treatment of cancer. Expert Opin. Drug Deliv.Opin. Drug Deliv. **18**(9), 1261–1290 (2021). https://doi.org/10.1080/17425247.2021.1912008
13. Dheyab, M.A., Aziz, A.A., Oladzadabbasabadi, N., Alsaedi, A., Braim, F.S., Jameel, M.S., Ramizy, A., Alrosan, M., Almajwal, A.M.: Comparative analysis of stable gold nanoparticles synthesized using sonochemical and reduction methods for antibacterial activity. Mol. **28**(9), 3931 (2023). https://doi.org/10.3390/molecules28093931
14. Freitas de Freitas, L., Varca, G.H.C., Dos Santos Batista, J.G., Benévolo Lugão, A.: An overview of the synthesis of gold nanoparticles using radiation technologies. Nanomaterials **8**(11), 939 (2018). https://doi.org/10.3390/nano8110939
15. Fritea, L., Banica, F., Costea, T.O., Moldovan, L., Dobjanschi, L., Muresan, M., Cavalu, S.: Metal nanoparticles and carbon-based nanomaterials for improved performances of electrochemical (Bio) sensors with biomedical applications. Materials **14**(21), 6319 (2021). https://doi.org/10.3390/ma14216319
16. Fuentes-García, J.A., Santoyo-Salzar, J., Rangel-Cortes, E., Goya, G.F., Cardozo-Mata, V., Pescador-Rojas, J.A.: Effect of ultrasonic irradiation power on sonochemical synthesis of gold nanoparticles. Ultrason. Sonochem. **70**, 105274 (2021). https://doi.org/10.1016/j.ultsonch.2020.105274
17. Hang, Y., Wang, A., Wu, N.: Plasmonic silver and gold nanoparticles: Shape- and structure-modulated plasmonic functionality for point-of-care sensing, bio-imaging, and medical therapy. Chem. Soc. Rev. **53** (2024). https://doi.org/10.1039/D3CS00793F
18. He, Q., Fang, Z., Shoshanim, O., Brown, S.S., Rudich, Y.: Scattering and absorption cross sections of atmospheric gases in the ultraviolet–visible wavelength range (307–725 nm). Atmos. Chem. Phys. **21**(19), 14927–14940 (2021). https://doi.org/10.5194/acp-21-14927-2021
19. Jangi, S. R. H., & Akhond, M.: Evaluation of biochemical behavior and stability of gold nanoparticles with high intrinsic peroxidase-like activity. Pet. Chem. Ind. Int. 6(4), 234–239 (2021). https://doi.org/10.33140/PCII.06.04.02
20. Kerry, R.G., Ukhurebor, K.E., Kumari, S., Maurya, G.K., Patra, S., Panigrahi, B., Patra, J.K.: A comprehensive review on the applications of nano-biosensor-based approaches for non-communicable and communicable disease detection. Biomater. Sci. **9**(10), 3576–3602 (2021). https://doi.org/10.1039/d0bm02164d
21. Khlebtsov, N.G.: Extinction, absorption, and scattering of light by plasmonic spheres embedded in an absorbing host medium. Phys. Chem. Chem. Phys. (2021). https://doi.org/10.1039/D1CP03057D
22. Khurana, K., Jaggi, N.: Localized surface plasmonic properties of Au and Ag nanoparticles for sensors: a review. Plasmonics **16**, 981–999 (2021). https://doi.org/10.1007/s11468-021-01381-1
23. Krishnan, S.: Colorimetric visual sensors for point-of-needs testing. Sens. Actuators Rep. (2022). https://doi.org/10.1016/j.snr.2022.100078
24. Kus-Liśkiewicz, M., Fickers, P., Ben Tahar, I.: Biocompatibility and cytotoxicity of gold nanoparticles: Recent advances in methodologies and regulations. Int. J. Mol. Sci. **22**(20), 10952 (2021). https://doi.org/10.3390/ijms222010952
25. Lalegani, Z., Ebrahimi, S.S., Hamawandi, B., La Spada, L., Batili, H., Toprak, M.S.: Targeted dielectric coating of silver nanoparticles with silica to manipulate optical properties for metasurface applications. Mater. Chem. Phys. **287**, 126250 (2022). https://doi.org/10.1016/j.matchemphys.2022.126250

26. Lee, S.A., Link, S.: Chemical interface damping of surface plasmon resonances. Acc. Chem. Res. (2021). https://doi.org/10.1021/acs.accounts.0c00872
27. Li, H., Li, J., Zhang, Y., Wang, X.: Gold nanoparticles combined with human β-defensin 3 gene-modified periodontal ligament cells promote osteogenic differentiation and periodontal regeneration via the p38 MAPK pathway. Front. Bioeng. Biotechnol. **9**, 631191 (2021). https://doi.org/10.3389/fbioe.2021.631191
28. Majdinasab, M., Daneshi, M., Marty, J.L.: Recent developments in non-enzymatic (bio) sensors for detection of pesticide residues: Focusing on antibody, aptamer and molecularly imprinted polymer. Talanta **232**, 122397 (2021). https://doi.org/10.1016/j.talanta.2021.122397
29. Montes-García, V., Squillaci, M.A., Diez-Castellnou, M., Ong, Q.K., Stellacci, F., Samori, P.: Chemical sensing with Au and Ag nanoparticles. Chem. Soc. Rev. **50**(2), 1269–1304 (2021). https://doi.org/10.1039/D0CS01112F
30. Nilghaz, A., Choi, J.R., Shen, W.: Noble-metal nanoparticle-based colorimetric diagnostic assays for point-ofneed applications. ACS Appl. Nano Mater. **4**(11), 12808–12824 (2021). https://doi.org/10.1021/acsanm.1c01545
31. Palani, S., Kenison, J.P., Sabuncu, S., Huang, T., Civitci, F., Esener, S., Nan, X.: Multispectral localized surface plasmon resonance (msLSPR) reveals and overcomes spectral and sensing heterogeneities of single gold nanoparticles. ACS Nano **17**(3), 2266–2278 (2023). https://doi.org/10.1021/acsnano.2c08702
32. Philip, A., Kumar, A.R.: The performance enhancement of surface plasmon resonance optical sensors using nanomaterials: a review. Coord. Chem. Rev.. Chem. Rev. (2022). https://doi.org/10.1016/j.ccr.2022.214424
33. Retout, M., Cornelio, B., Bruylants, G., Jabin, I.: Bifunctional calix[4]arene-coated gold nanoparticles for orthogonal conjugation. Langmuir, **38**(30), 9301–9309 (2022). https://doi.org/10.1021/acs.langmuir.2c01122
34. Sadiq, S., Khan, S., Khan, I., Khan, A., Humayun, M., Wu, P., Usman, M., Khan, A., Alanazi, A.F., Bououdina, M.: A critical review on metal-organic frameworks (MOFs) based nanomaterials for biomedical applications: Designing, recent trends, challenges, and prospects. Heliyon **10**(2), e25521 (2024). https://doi.org/10.1016/j.heliyon.2024.e25521
35. Sadiq, Z., Safiabadi Tali, S.H., Hajimiri, H., Al-Kassawneh, M., Jahanshahi-Anbuhi, S.: Gold nanoparticles-based colorimetric assays for environmental monitoring and food safety evaluation. Crit. Rev. Anal. Chem. **54**(7), 2209–2244 (2024). https://doi.org/10.1080/10408347.2022.2162331
36. Sahu, B., Kurrey, R., Deb, M. K., Shrivas, K., Karbhal, I., Khalkho, B.R.: A simple and cost-effective paperbased and colorimetric dual-mode detection of arsenic(III) and lead(II) based on glucose-functionalized gold nanoparticles. RSC Adv. **11**(33), 20769–20780 (2021). https://doi.org/10.1039/D1RA02929K
37. Singh, H., Bamrah, A., Bhardwaj, S.K., Deep, A., Khatri, M., Brown, R.J., Kim, K.H.: Recent advances in the application of noble metal nanoparticles in colorimetric sensors for lead ions. Environ. Sci. Nano **8**(4), 863–889 (2021). https://doi.org/10.1039/D0EN00963F
38. Tang, Z., Wu, J., Yu, X., Hong, R., Zu, X., Lin, X., Yi, G.: Fabrication of Au nanoparticle arrays on flexible substrate for tunable localized surface plasmon resonance. ACS Appl. Mater. Interfaces **13**(7), 9281–9288 (2021). https://doi.org/10.1021/acsami.0c22785
39. Tayang, A., Songachan, L.S.: Microbial bioremediation of heavy metals. Curr. Sci (2021). https://doi.org/10.18520/cs/v120/i6/1013-1025
40. Umapathi, R., Park, B., Sonwal, S., Rani, G.M., Cho, Y., Huh, Y.S.: Advances in optical-sensing strategies for the on-site detection of pesticides in agricultural foods. Trends Food Sci. Technol. **119**, 69–89 (2022). https://doi.org/10.1016/j.tifs.2021.11.018
41. Umapathi, R., Sonwal, S., Lee, M.J., Rani, G.M., Lee, E.S., Jeon, T.J., Huh, Y.S.: Colorimetric based on-site sensing strategies for the rapid detection of pesticides in agricultural foods: new horizons, perspectives, and challenges. Coord. Chem. Rev. Chem. Rev. **446**, 214061 (2021). https://doi.org/10.1016/j.ccr.2021.214061
42. Yu, X., Wang, Y., Chen, Z.: Gold nanoparticles: Synthesis, physiochemical properties and therapeutic applications in cancer. Drug Discovery Today, **26**(3), 735–745 (2021). https://doi.org/10.1016/j.drudis.2021.01.030

Chapter 3
Copper Nanoparticles in Colorimetric Sensing

Abstract Copper nanoparticles (CuNPs) have been extensively utilized as a material for colorimetric sensing due to their size-dependent unique plasmonic properties, low-cost, and catalytic activity. These nanoparticles demonstrate a surface plasmon resonance (SPR) peak in the visible-range of ~ 530 nm that can be finely adjusted by particle size and morphology. Owing to their high surface-to-volume ratio, CuNPs exhibit enhanced interaction with target analytes, which boosts their catalytic activity and amplifies the colorimetric signal, resulting in improved sensitivity. These properties enable their use in diverse applications, including biomolecule detection, environmental monitoring, and medical diagnostics. Still, oxidation and stability are the major challenges that need to be mitigated. The performance and reproducibility of CuNPs can alleviate these challenges by surface functionalization and the fabrication of composite structures. In this chapter, we specifically discuss CuNP-based colorimetric sensors, including their properties, reaction mechanisms, and practical applications, and emphasize their potential in real world applications. Finally, it discusses the remaining challenges and attempts to anticipate the future trends driven at improving the sensitivity, selectivity and stability of such metal nanoparticle sensors.

Copper nanoparticles (CuNPs) are of great interest in nanotechnology due to their low cost, visible range plasmonic properties, and good catalytic activity for sensing and driving reactions [19, 34]. CuNPs are easily modified for practical application and, thus exhibit high potential as a sensing platform. Over the past decade, various advanced materials based on CuNPs have been developed, demonstrating significantly improved performance in colorimetric sensing applications compared to earlier generations. Over the past three years, there have been 35 new CuNP-based colorimetric sensing materials that have been introduced [28, 29, 40]. These CuNPs can serve multiple purposes, including antibacterial and antiviral activities, catalysis, photocatalysis, and cancer imaging, due to their oxidation and physical properties [30] as shown in Fig. 3.1.

A. C. Alguno et al., *Metal Nanoparticles-Based Nanoplatforms for Colorimetric Sensing*, Engineering Materials, https://doi.org/10.1007/978-981-96-6771-0_3

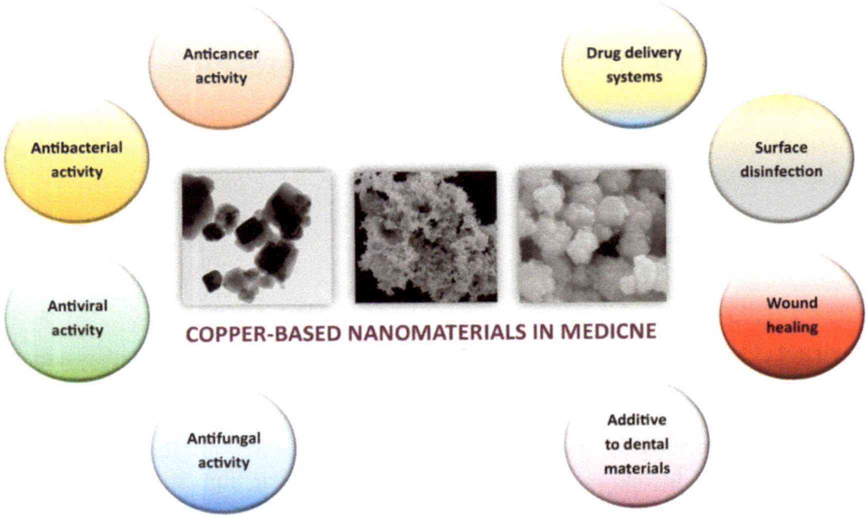

Fig. 3.1 The characteristics, properties and various applications of CuNPs. Adapted from [30], Fig. 3.1. Licensed under CC BY 4.0 (https://creativecommons.org/licenses/by/4.0/)

3.1 Unique Properties of Copper Nanoparticles

Considering their properties, CuNPs are gradually being developed for colorimetric sensing [2]. CuNPs can be used to detect a wide range of analytes across various fields, as their properties often surpass those of other metal nanoparticles. Their high surface-to-volume ratio increases their catalytic activity, which in turn increases their probability of interacting with the analytes [8, 25]. Having transitioned to the nanoscale, CuNPs are more selective and sensitive than their bulk counterparts. Additionally, their size and morphology can be tailored to enhance their effectiveness in a range of biological and chemical sensing applications[18].

Given their excellent optical properties, CuNPs exhibit color changes based on parameters such as particle size, dielectric properties, and the surrounding optical fields [14, 24]. The surface plasmon resonance (SPR) of CuNPs can be finely tuned and controlled by size and shape and their scattering peak is around 530 nm which has appreciable absorbance unlike previously reported nanostructures around 450 nm. A further important factor is that CuNPs display good antibacterial properties, suggesting that these NPs are biocompatible and stable under distinct conditions [27, 30]. CuNPs act in agglomeration with other materials to form composite structures that allow improved properties for sensor applications while improving sensitivity in experiments. CuNPs can be synthesized using various methods, and their color varies depending on the biosynthesis approach. Typically, enzymatic biosynthesis involves functional groups in biomolecules, resulting in a relatively narrow size distribution, an important factor for their use in a range of scientific applications.

3.1.1 *High Surface Area and Catalytic Activity*

The large surface area of CuNPs enhances their catalytic activity, making them highly effective for interacting with a wide range of target analytes [26, 35]. CuNPs smaller than 10 nm are more reactive due to their high surface to volume ratio and numerous concave edges. This catalytic magnification leads to significantly enhanced colorimetric signals for sensing applications. The larger the surface area, the higher the capacity for adsorbing and retaining the target analytes on CuNPs which consequently enhances their colorimetric sensitivity. Further, reducing CuNP size has also been shown to significantly enhance the limit of detection (LOD) for oxytetracycline, from 4.4 to 1.33 nm when CuNP size was reduced from 20.2 to 5.2 nm [13, 22]. In addition, the LOD of 7.88 fM and linear detection range of 0.01–500 pM for Al (III) were obtained with very good results. As a result, numerous CuNP-based colorimetric platforms have been developed for detecting various target analytes.

Over the years, CuNPs with excellent properties were prepared for various applications. CuNPs play a vital role not only in signal amplification for the detection of metal ions and bacteria but also in enhancing colorimetric signals generated by biochemical and chemical reactions [28, 29, 40]. This chapter presents an overview of various types of CuNPs and their characteristics, emphasizing the impact of surface area and reaction mechanisms on signal amplification. It also reviews the practical applications of CuNPs and evaluates their potential in guiding the development of innovative sensing strategies in the future.

3.2 Mechanistic Insights into CuNPs-Based Colorimetric Reactions

CuNPs have been of great interest due to their optical properties and are very representative of the family of colorimetric metallic nanoparticles [11]. They have improved redox responses when combined with different analytes, and they can also play a catalytic role in forming the analytes and their chemical structure changes. This changes the color of the solution to a darker or brighter color that is just visible to the naked eyes, or quantifiable by UV–vis spectrophotometry with the ultimate goal of improving colorimetric sensing technology. The unique catalytic activity of CuNPs makes them highly selective toward specific analytes, a key feature in electrocatalysis. Certain analytes can increase the electrochemical potential, promoting the formation of more Cu_2O in proportion to the analyte present. This highlights the significance of CuNPs in analytical applications, where they are used to enable precise and reliable detection methods.

The sensing mechanisms of electron transfer can be divided into two types, i.e., signal-off sensors and signal-on sensors, which are characterized by the darkening and brightening of the solutions, respectively [20, 36]. The "turn-off" response is characterized by a shift in the LSP band from blue to red, while the "turn-on" response involves a red-to-blue transition. The first section emphasizes the importance of reaction rates in sensing processes and the relevance of understanding reaction mechanisms. It also discusses key factors affecting detection efficiency, such as nanoparticle size and surface characteristics. Additionally, the chapter will present a referenced study focused on the use of CuNPs, along with practical applications of modern colorimetric methods. Gaining a deep understanding of these detection mechanisms is essential, as it forms the basis for designing next-generation CuNP-based reagents. These advancements aim to broaden their applicability and significantly enhance detection performance.

This has focused the attention of the scientists towards the nature of these probes especially the CuNPs which proves useful as a colorimetric sensor in environment for various analytes [14, 37]. However, the rationale behind the selection of certain analytes over others remains an unresolved challenge. This chapter aims to provide a comprehensive overview of CuNP-based colorimetric detection, focusing on its analytical parameters and specifications, as well as highlighting the advantages and limitations of these methods.

3.2.1 Redox Reactions and Interaction with Various Analytes

A redox reaction occurs between Cu and a targeted analyte, that is either an ion, biomolecule, or drug [16, 17] to enhance sensitivity in colorimetric sensing techniques. Copper ions and *in situ*-formed nanozymes can be detected using methods like Inductively Coupled Plasma Mass Spectrometry and Resonance Light Scattering. The interaction of various biomolecules with CuNPs plays a critical role in the operation of these analytical devices. In this work, multiple pathways for CuNP oxidation and the resulting colorimetric changes will be explored in detail.

3.2.2 Effect of Size and Surface Modifications on Detection Efficiency

The synthesis processes and size attributes of metal nanoparticles, particularly CuNPs, significantly influence their optical properties, surface area, reactive site quantity, height, and catalytic activity [3, 5]. These size-dependent properties highlight the unique potential of CuNPs in advanced microscale analytical technologies. Well-defined CuNPs exhibit LSPR, enabling interaction with a broad spectrum of light wavelengths. However, their sharp edges can increase susceptibility to oxidation, potentially affecting both stability and overall performance.

Different methods including covalent and physical binding are used to successfully immobilize recognition elements on metal nanoparticle surfaces [1, 10].

Depending on the interaction mechanisms involved, recognition elements may either remain bound to or detach from the nanoparticles – a process that is critical for capturing target analytes. Colorimetric methods are particularly valuable for detecting biological and environmental targets, even at low concentrations. However, limited research has been conducted on how the size of CuNPs affects the performance of colorimetric assays. Studies have shown that tuning nanoparticle size and surface properties can significantly enhance their reactivity and affinity toward key analytes. Therefore, optimizing these parameters is essential to achieve high sensitivity and selectivity in analytical applications. Nonetheless, challenges related to nanoparticle toxicity and stability continue to restrict their broader practical use.

3.3 Specific Applications of Copper Nanoparticle Sensors

CuNPs are used to detect dopamine with high selectivity, as the presence of other compounds such as cysteine, L-ascorbic acid, and uric acid does not affect the color of the nanoparticle solution [6, 12]. The sensor exhibits high specificity for dopamine, primarily due to the strong affinity of CuNPs for binding selectively to dopamine over other substances. This selectivity is further enhanced by the catalytic activity of CuNPs in facilitating the formation of a borate–dopamine chelation complex [4, 38]. Dissolved oxygen promotes the oxidation of the borate–dopamine complex by hydrogen peroxide, resulting in a visible color change from brown to dark blue. This is accompanied by a linear increase in optical absorbance with rising dopamine concentrations in the range of 0.02–10.0 mM, with a low detection limit of 0.0155 μM. This method is cost-effective, and the sensor is both reusable and efficient for determining dopamine concentrations across various sample matrices.

3.3.1 Detection of Biomolecules

Sensing of biological matters, especially proteins and enzymes, has become an increasingly identified area of investigation within the past few years [7]. MNPs are suitable for colorimetric sensing applications because the aggregation of MNPs due to biomolecule interactions can be observed through a difference in color. In addition, the flower-like or sphere-like type nanoparticles have performances that are effective in detecting a range of proteins.

Moreover, one brilliant group has constructed a sensitive surface-enhanced Raman scattering (SERS) active aptasensor based upon site-specific protein aggregated nanoparticles which could detect target proteins such as platelet-derived growth factor and human immunoglobulin G with the assistance of site-selective induced arrangement without sample pretreatment in real-time, suitable for rapid biological analysis [32, 41]. To enable UV–vis absorption signal changes for colorimetric detection, these systems utilize the nanoparticle aggregation mechanism, which produces a distinct and visible color shift. A more sensitive assay strategy (detection limit = 0.24 μg mL^{-1} in the broad range of 0.5–150 μg mL^{-1}) was also reported.

3.3.2 Monitoring of Environmental Pollutants

The high extinction coefficients, unique electronic and optical properties, and exceptionally high surface-to-volume ratios of metal nanoparticle-based nanoplatforms enable them to be favorable signal transducers making them effective colorimetric probes for the analysis of reversible binding of analytes with their receptor, encompassing both biomolecules and chemical molecules. Metal nanoparticles are considered as basic building blocks of the nanoplatforms to develop and enhance the performance of colorimetric sensor assays.

The detection of environmental contaminants, including organic, inorganic, and biological substances, has gained significant attention due to their harmful effects on living organisms and the environment [33]. This issue is widespread, highlighting the need for affordable, simple, and reliable colorimetric sensors. Traditional analytical systems tend to be large, complex, expensive, and not very practical. However, advancements in nanomaterials and colorimetry have led to the development of metal nanoparticle-based sensors that are simple, rapid, sensitive, and selective. These sensors have demonstrated significant effectiveness in detecting various analytes. The colorimetric responses, such as shifts towards red or blue or the formation of aggregates, result from the interaction mechanisms between different species. Shown in Fig. 3.2 is the detection of Hg^{2+} using surface-enhanced Raman spectroscopy (SERS) active mercury ions interact with SERS-active molecular recognition elements (MREs) via three mechanisms: (a) electrophilic substitution between Hg^{2+} and 4-mercaptophenylboronic acid (4-MPBA), (b) reusable 4-mercaptopyridine (4-MPy)/AuNPs/indium tin oxide (ITO) chips for Hg^{2+} detection, and (c) a SERS frequency-shift method to distinguish Hg^{2+} from $MeHg^{+}$ via dimethyldithiocarbamic acid sodium salt (DASS) binding.

Furthermore, the application of nanomaterials in colorimetric sensor assays offers vast opportunities for noble metal nanoparticles, which include Au, Ag, and binary polynanocomposites. With the increasing recognition of the advantages of both colorimetry and noble metal nanoparticles, their applications in detecting environmental pollutants will also expand. Nevertheless, challenges like the combination of colorimetric sensors with other analytical methods and the monitoring of pollutants *in vivo* in real samples remain [9].

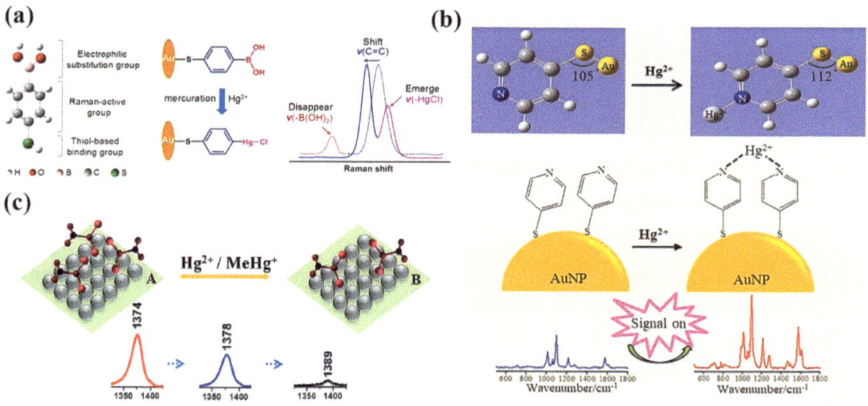

Fig. 3.2 Detection of Hg^{2+} using surface-enhanced Raman spectroscopy (SERS) active mercury ions interact with SERS-active molecular recognition elements (MREs) via interaction with Hg^{2+} ions **a** Recognition of Hg^{2+} ion through an electrophilic substitution reaction between Hg^{2+} and 4-mercaptophenylboronic acid (4-MPBA). **b** Reusable 4-mercaptopyridine (4-MPy)/AuNPs/indium tin oxide (ITO) chips for Hg^{2+} ion sensing. **c** A frequency-shift based SERS approach for discriminating between Hg^{2+} and $MeHg^{+}$ through binding with dimethyldithiocarbamic acid sodium salt (DASS). Adapted from [31], Fig. 3.2. Licensed under CC BY 4.0 (https://creativecommons.org/licenses/by/4.0/)

3.4 Challenges in the Use of Copper Nanoparticles

CuNPs being conductive, thermally stable, and having chemical resistance have been beneficial in wide area of research and applications [15, 28, 29]. Nonetheless, the application of CuNPs in colorimetric sensors faces considerable challenges. In the synthesis process, Cu^{2+} in oxide and borate forms would be usually exist, which are not stable nor active. In addition, there is no systematic framework that explains how the analytes modify the color of the system, which makes it difficult to develop high-fidelity sensors. Furthermore, the stability and reproducibility of CuNP-based systems are already challenging, and the presence of oxygen pollutants exacerbates this issue.

CuNPs are cost-effective due to the abundance of copper, making them more affordable than AuNPs and AgNPs. However, their high reactivity leads to lower stability, as they are easily oxidized, which can reduce their sensitivity and reliability in long-term sensing applications. In contrast, AuNPs offer enhanced optical and plasmonic properties, making them more sensitive, though they come at a higher cost. AgNPs, while even more sensitive, are toxic compared to CuNPs. To enhance the sensitivity of CuNPs without increasing costs, strategies such as surface stabilization through functionalization or encapsulation, as well as the development of green synthesis methods, are essential. Addressing these challenges is crucial for ensuring the long-term effectiveness of CuNPs in colorimetric sensing and other applications.

3.4.1 Issues Related to Oxidation, Stability, and Reproducibility

In their normal environment, nanomaterials when subjected to oxidation or irradiation develop an oxide shell that deleteriously effects their LSPR signals, an essential property for optimizing sensing applications. This performance limitation has spurred interest in multilayer nanocomposites to improve oxidation resistance. The hydrophobicity of these advanced composites can be further adjusted to incorporate a wide range of thiolated capture ligands, expanding the spectrum of detectable entities. However, the instability of thiol-metal interactions can result in unreliable sensors. To address this, structures can be designed with controlled slide radii to enhance both stability and reproducibility, ensuring longer-lasting, reliable sensors. On one hand, stable citrate-coated nanoparticles improve the shelf life of reagents with their long-lasting stability and reproducible outputs, which is crucial for practical and laboratory applications. Additionally, the size-stability of such nanoparticles enables LSPR analysis for real-time or point-of-sample diagnostics. While AuNPs offer the best stability and optical efficiency, they come with a high price tag. AgNPs, on the other hand, are highly sensitive and cost-effective but suffer from toxicity and instability. CuNPs are more affordable but are highly prone to oxidation. Moreover, there is a need to improve the sustainability of nanoparticle synthesis and address the lack of proper stabilization techniques to ensure long-term effectiveness.

References

1. Ahmad, F., Salem-Bekhit, M.M., Khan, F., Alshehri, S., Khan, A., Ghoneim, M.M., Elbagory, I.: Unique properties of surface-functionalized nanoparticles for bio-application: functionalization mechanisms and importance in application. Nanomaterials **12**(8), 1333 (2022). https://doi.org/10.3390/nano12081333
2. Ali, Z., Ullah, R., Tuzen, M., Ullah, S., Rahim, A., Saleh, T.A.: Colorimetric sensing of heavy metals on metal doped metal oxide nanocomposites: a review. Trends Environ. Anal. Chem. **37**, e00187 (2023). https://doi.org/10.1016/j.teac.2022.e00187
3. Behzad, F., Naghib, S.M., Tabatabaei, S.N., Zare, Y., Rhee, K.Y.: An overview of the plant-mediated green synthesis of noble metal nanoparticles for antibacterial applications. J. Ind. Eng. Chem. **94**, 92–104 (2021). https://doi.org/10.1016/j.jiec.2020.12.005
4. Cao, W., Guo, T., Wang, J., Xu, G., Jiang, J., Liu, D.: Cu-based materials: design strategies (hollow, core-shell, and LDH), sensing performance optimization, and applications in small molecule detection. Coord. Chem. Rev.. Chem. Rev. **497**, 215450 (2023). https://doi.org/10.1016/j.ccr.2023.215450
5. Chandrakala, V., Aruna, V., Angajala, G.: Review on metal nanoparticles as nanocarriers: current challenges and perspectives in drug delivery systems. Emergent Mater. (2022). https://doi.org/10.1007/s42247-021-00335-x

6. Chung, K., Bang, J., Thacharon, A., Song, H.Y., Kang, S.H., Jang, W.S., Kim, S.W.: Non-oxidized bare copper nanoparticles with surface excess electrons in air. Nat. Nanotechnol.Nanotechnol. **17**(3), 285–291 (2022). https://doi.org/10.1038/s41565-021-01070-4
7. Dawson, K.A., Yan, Y.: Current understanding of biological identity at the nanoscale and future prospects. Nat. Nanotechnol.Nanotechnol. (2021). https://doi.org/10.1038/s41565-021-00860-0
8. Dinu, L.A., Kurbanoglu, S., Romanitan, C., Pruneanu, S., Serban, A.B., Stoian, M.C., Craciun, G.: Electrodeposited copper nanocubes on multi-layer graphene: a novel nanozyme for ultra-sensitive dopamine detection from biological samples. Appl. Surf. Sci. **604**, 154392 (2022). https://doi.org/10.2139/ssrn.4128534
9. Fritea, L., Banica, F., Costea, T.O., Moldovan, L., Dobjanschi, L., Muresan, M., Cavalu, S.: Metal nanoparticles and carbon-based nanomaterials for improved performances of electrochemical (Bio) sensors with biomedical applications. Materials **14**(21), 6319 (2021). https://doi.org/10.3390/ma14216319
10. Gao, S., Guisán, J.M., Rocha-Martin, J.: Oriented immobilization of antibodies onto sensing platforms—a critical review. Anal. Chim. ActaChim. Acta (2022). https://doi.org/10.1016/j.aca.2021.338907
11. Gerdan, Z., Saylan, Y., Denizli, A.: Recent advances of optical sensors for copper ion detection. Micromachines (2022). https://doi.org/10.3390/mi13081298
12. Godiya, C.B., Kumar, S., Park, B.J.: Superior catalytic reduction of methylene blue and 4-nitrophenol by copper nanoparticles-templated chitosan nanocatalyst. Carbohydr. Polym. Technol. Appl. **5**, 100267 (2023). https://doi.org/10.1016/j.carpta.2022.100267
13. Haider, F., Mashrafi, M., Aoni, R.A., Haider, R., Hossen, M., Ahmed, T., Ahmed, R.: Multi-analyte detection based on integrated internal and external sensing approach. IEEE Trans. Nanobiosci.Nanobiosci. **21**(1), 29–36 (2021). https://doi.org/10.1109/TNB.2021.3108834
14. Hyder, A., Buledi, J.A., Nawaz, M., Rajpar, D.B., Orooji, Y., Yola, M.L., Solangi, A.R.: Identification of heavy metal ions from aqueous environment through gold, silver and copper nanoparticles: an excellent colorimetric approach. Environ. Res. **205**, 112475 (2022). https://doi.org/10.1016/j.envres.2021.112475
15. Kumar, J.A., Prakash, P., Krithiga, T., Amarnath, D.J., Premkumar, J., Rajamohan, N., Rajasimman, M.: Methods of synthesis, characteristics, and environmental applications of MXene: a comprehensive review. Chemosphere **286**, 131607 (2022). https://doi.org/10.1016/j.chemosphere.2021.131607
16. Jiang, Y., Huo, Z., Qi, X., Zuo, T., Wu, Z.: Copper-induced tumor cell death mechanisms and antitumor theragnostic applications of copper complexes. Nanomedicine (2022). https://doi.org/10.2217/nnm-2021-0374
17. Jing, G., Yang, L., Wang, H., Niu, J., Li, Y., Wang, S.: Interference of layered double hydroxide nanoparticles with pathways for biomedical applications. Adv. Drug Deliv. Rev.Deliv. Rev. **188**, 114451 (2022). https://doi.org/10.1016/j.addr.2022.114451
18. Manjula, N.G., Sarma, G., Shilpa, B.M., Suresh Kumar, K.: Environmental applications of green engineered copper nanoparticles. In: Phytonanotechnology, pp. 255–276. Springer Nature Singapore, Singapore (2022). https://doi.org/10.1007/978-981-19-4811-4_12.
19. Li, X., Zhu, H., Liu, P., Wang, M., Pan, J., Qiu, F., Niu, X.: Realizing selective detection with nanozymes: strategies and trends. TrAC, Trends Anal. Chem. **143**, 116379 (2021). https://doi.org/10.1016/j.trac.2021.116379
20. Liu, Q., Zhang, H., Jiang, H., Yang, P., Luo, L., Niu, Q., You, T.: Photoactivities regulating of inorganic semiconductors and their applications in photoelectrochemical sensors for antibiotics analysis: a systematic review. Biosens. Bioelectron.. Bioelectron. **216**, 114634 (2022). https://doi.org/10.1016/j.bios.2022.114634
21. Majumdar, D., Ghosh, S.: Recent advancements of copper oxide based nanomaterials for supercapacitor applications. J. Energy Storage (2021). https://doi.org/10.1016/j.est.2020.101995

22. Min, H., Chen, Z., Han, Z., Wang, K., Xu, J., Shi, W., Cheng, P.: Detection of the UV-vis silent biomarker trimethylamine-N-oxide via outer-sphere interactions in a lanthanide metal-organic framework. Commun. Chem. **5**(1), 74 (2022). https://doi.org/10.1038/s42004-022-00690-8
23. Navas, D., Fuentes, S., Castro-Alvarez, A., Chavez-Angel, E.: Review on sol-gel synthesis of perovskite and oxide nanomaterials. Gels **7**(4), 275 (2021). https://doi.org/10.3390/gels7040275
24. Sami, A.S., Barani, S.S.I., Rashid, R.F., Zeebaree, S.Y.S., Zeebaree, A.Y.S., Zebari, O.I.H., Issa, M.M.: Novel biosensor for highly sensitive detection of serum albumin in artificial human urine using CuNPs@ AG. Sens. Bio-Sensing Res. **43**, 100633 (2024). https://doi.org/10.1016/j.sbsr.2024.100633
25. Thao, B.T.P., Van Trang, N., Le Huy, N., Dung, N.Q., Ahmed, M.U., Dai Lam, T.: The role of copper nanoparticles decorating polydopamine/graphene film as catalyst in the enhancement of uric acid sensing. J. Electroanal. Chem.Electroanal. Chem. **893**, 115322 (2021). https://doi.org/10.1016/j.jelechem.2021.115322
26. Truong, T.T., Mondal, S., Doan, V.H.M., Tak, S., Choi, J., Oh, H., Oh, J.: Precision-engineered metal and metal-oxide nanoparticles for biomedical imaging and healthcare applications. Adv. Colloid Interface Sci. 103263 (2024). https://doi.org/10.1016/j.cis.2024.103263
27. Vodyashkin, A., Stoinova, A., Kezimana, P.: Promising biomedical systems based on copper nanoparticles: synthesis, characterization, and applications. Colloids Surf. B: Biointerfaces 113861 (2024). https://doi.org/10.1016/j.colsurfb.2024.113861
28. Wang, H., Wu, T., Li, M., Tao, Y.: Recent advances in nanomaterials for colorimetric cancer detection. J. Mater. Chem. B (2021). https://doi.org/10.1039/D0TB02163F
29. Wang, X., Zhang, Y., Wu, J., Zhang, Z., Liao, Q., Kang, Z., Zhang, Y.: Single-atom engineering to ignite 2D transition metal dichalcogenide based catalysis: fundamentals, progress, and beyond. Chem. Rev. **122**(1), 1273–1348 (2021). https://doi.org/10.1021/acs.chemrev.1c00505
30. Woźniak-Budych, M.J., Staszak, K., Staszak, M.: Copper and copper-based nanoparticles in medicine—perspectives and challenges. Molecules (2023). https://doi.org/10.3390/molecules28186687
31. Xu, G., Song, P., Xia, L.: Examples in the detection of heavy metal ions based on surface-enhanced Raman scattering spectroscopy. Nanophotonics **10**(18): 4419–4445 (2021). https://doi.org/10.1515/nanoph-2021-0363
32. Yang, Z., Guo, J., Wang, L., Zhang, J., Ding, L., Liu, H., Yu, X.: Nanozyme-enhanced electrochemical biosensors: mechanisms and applications. Small (2024). https://doi.org/10.1002/smll.202307815
33. Yang, G.L., Jiang, X.L., Xu, H., Zhao, B.: Applications of MOFs as luminescent sensors for environmental pollutants. Small (2021). https://doi.org/10.1002/smll.202005327
34. Ye, M.L., Zhu, Y., Lu, Y., Gan, L., Zhang, Y., Zhao, Y.G.: Magnetic nanomaterials with unique nanozymes-like characteristics for colorimetric sensors: a review. Talanta (2021). https://doi.org/10.1016/j.talanta.2021.122299
35. You, Y., Deng, Q., Wang, Y., Sang, Y., Li, G., Pu, F., Qu, X.: DNA-based platform for efficient and precisely targeted bioorthogonal catalysis in living systems. Nat. Commun.Commun. **13**(1), 1459 (2022). https://doi.org/10.1038/s41467-022-29167-x
36. Zhang, X., Wang, F., Zhi, H., Zhao, J., Wan, P., Feng, L.: Electrochemical "signal on/off" paper-based aptasensor for ochratoxin a detection based on MXene-Au and Pt@ NiCo-LDH-catalyzed signal amplification. Sens. Actuators, B Chem. **368**, 132161 (2022). https://doi.org/10.2139/ssrn.4076754
37. Zeebaree, A.Y.S., Zebari, O.I.H., Zeebaree, S.Y.S., Thanoon, R.D., Yaseen, Y.N.: Current contributions of novel nanoparticles-based colorimetric sensors for detection of SCN ions in different aqueous models. A review. Arab. J. Chem. **16**(12), 105297 (2023). https://doi.org/10.1016/j.arabjc.2023.105297
38. Zhang, L., Wang, H., Qu, X.: Biosystem-inspired engineering of nanozymes for biomedical applications. Adv. Mater. (2024). https://doi.org/10.1002/adma.202211147

39. Zhao, X., Li, Z., Ding, Z., Wang, S., Lu, Y.: Ultrathin porous Pd metallene as highly efficient oxidase mimics for colorimetric analysis. J. Colloid Interface Sci. **626**, 296–304 (2022). https://doi.org/10.1016/j.jcis.2022.06.124
40. Zhou, Y., Huang, X., Hu, X., Tong, W., Leng, Y., Xiong, Y.: Recent advances in colorimetry/fluorimetry-based dual-modal sensing technologies. Biosens. Bioelectron.. Bioelectron. **190**, 113386 (2021). https://doi.org/10.1016/j.bios.2021.113386
41. Zhou, Z., Yang, Z., Xia, L., Zhang, H.: Construction of an enzyme-based all-fiber SPR biosensor for detection of enantiomers. Biosens. Bioelectron. (2022). https://doi.org/10.1016/j.bios.2021.113836

Chapter 4
Iron Nanoparticles for Colorimetric Sensing

Abstract Due to their distinctive properties such as high affordability, magnetic properties, and peroxidase mimetic catalytic activity, iron nanoparticles (FeNPs) have drawn great interest in the field of colorimetric sensing. Such nanoparticles show remarkable sensing properties for detecting various analytes such as biomolecules, neurotransmitters, ions, and environmental pollutants. This versatility is highlighted also in this chapter detailing the synthesis, properties and mechanisms of iron-based colorimetric platforms, as well as their subsequent use in the field of medical diagnostic and environmental monitoring. FeNPs, notably Fe_3O_4, have shown excellent catalytic properties that can imitate enzyme mimetics; these reactions can be employed for oxidation and reduction reactions whose products often produce easily observable colorimetric changes so that they can be used for qualitative and quantitative assessment. In addition, the chapter addresses emerging hybrid systems (e.g., Fe containing Metal–Organic Frameworks (MOFs)), which can facilitate selective recognition of the desired analyte, thereby improving sensors performance. While they do have advantages, issues of stability, specificity, and scalability persist, which has prevented their wider application. This chapter discusses these issues and gives future directions for enhancing the sensitivity, reproducibility and practicability of the iron-based colorimetric sensors.

4.1 Introduction to Iron Nanoparticles and Derivatives

Researchers have investigated iron nanoparticles (FeNPs) and their derivatives because of their distinctive physical and chemical characteristics [17]. As a result, FeNPs are extremely good candidate for applications in sensing and catalysis. FeNPs have optical features that allow them to detect specific analytes by colorimetric sensing. As these applications are very critical in the evolution of analytical chemistry for the systematic studies on diversified metal nanoparticles which played an important role in comprehending their properties and catalytic behaviors. Controllable and monodispersed FeNPs can be synthesized from soluble metal salts including

A. C. Alguno et al., *Metal Nanoparticles-Based Nanoplatforms for Colorimetric Sensing*, Engineering Materials, https://doi.org/10.1007/978-981-96-6771-0_4

iron chloride or iron acetate employing various methods to fine-tune materials properties [8, 21]. Different methods of preparation of nanoparticles are utilized to control the size and shape of nanoparticles for a specific application. The chemical environment and aggregation state of FeNPs determine their optical properties, which are unique among metallic systems [3, 16, 25].

To illustrate FeNPs, especially magnetic Fe_3O_4 nanoparticles, Fig. 4.1 is included. The aspect ratio and length of these nanorods can be tuned by changing the amount of precursors and the reaction time to form the nanostructures. The synthesized Fe_3O_4 nanorods were characterized and the surface features were displayed in the transmission electron microscopy (TEM) and high-resolution transmission electron microscopy (HR-TEM) images (Fig. 4.1) [24]. Furthermore, the image represents the structural diversity of Fe_3O_4 nanoparticles, which can manifest as nanorods, porous hollow nanoparticles, hexagonal nanoplates, tripods, and tetrapods depending on the synthesis method.

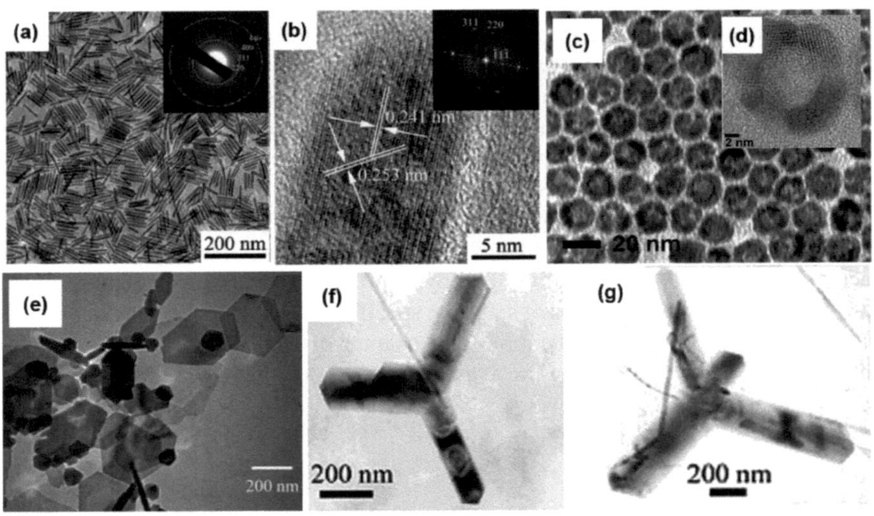

Fig. 4.1 **a, b** TEM and HR-TEM pictures of Fe_3O_4 nanorods. **c, d** TEM and HR-TEM images that show the structure of porous hollow nanoparticles (NPs). **e** A transmission electron microscope (TEM) image showing hexagonal nanoplates in two dimensions. **f, g** TEM images of Fe_3O_4 tripods and tetrapods. Adapted from [24], Fig. 4.1. Licensed under CC BY 4.0 (https://creativecommons.org/licenses/by/4.0/)

4.1.1 Peroxidase-Like Activities of Fe_3O_4 and Other Iron Oxides

Two important iron oxides, Fe_3O_4 (magnetite) and Fe_2O_3 (maghemite), exhibit peroxidase activity with environmental responses being well suited for catalyzing reactions under acidic conditions and low H_2O_2 concentration. Such oxides have a broader spectrum than the natural counterpart peroxidase [2, 5] and these oxides can resist enzyme inhibitors. The nanoscale size of these materials significantly enhances their catalytic efficiency, positioning them as promising tools across diverse scientific disciplines. These nanoparticles catalyze the oxidation of substrates into colorimetric detector, enabling the formation of immunological complexes and facilitating both qualitative and quantitative detection, thereby accelerating progress in biomedical research and diagnostics [14, 32]. The surface modification of Fe_3O_4 nanoparticles is significant in a way that it leads to the synthesis of Horseradish Peroxidase (HRP)-like nanozymes which are more catalytically efficient compared to natural HRP. The activity of these nanozymes can be modulated by controlling the size, surface decoration, and dopants [6, 22]. These nanozymes demonstrate strong catalytic activity and excellent stability, enabling H_2O_2-induced iron oxide precipitation upon analyte addition, which serves as a reliable basis for colorimetric detection. Furthermore, their enzymatic performance can be finely tuned through the incorporation of protective surface shells, allowing for precise modulation of activity under specific pH conditions.

4.2 Mechanisms of Iron-Based Colorimetric Platforms

Iron based colorimetric platforms are currently the most popular and are used for the determination of biomolecules, proteins, substances associated with diseases, neurotransmitters, ions, H_2O_2, GSH and bacteria [18, 36]. The formation of colorimetric changes by iron nanoparticles is done by chemical or biochemical synthesis and the optical features of the nanoparticles is a function of the surface to volume ratio as well as the catalytic activity towards the enhancement of the sensing signals. These platforms are also highly selective and sensitive in the analysis of target biomolecules when combined. Moreover, the residual metallic iron and its hybrids, which are produced via the reduction of metal ions during the synthesis process, are employed for loading target molecules [23, 34, 35]. This residual iron catalyzes the reaction of Tetramethylbenzidine (TMB) and H_2O_2 to bring about colorimetric changes from colorless to teal or blue. The iron nanoparticles in water are more hydrophilic and have better dispersity due to the formation of hydroxyl groups from iron oxides such as Fe_3O_4 or Fe_2O_3 and therefore these have great potential for application in scientific research and industrial diagnostics.

4.2.1 Enzyme Mimicking Behaviors and Catalytic Pathways

Metal nanoparticles (MNPs) exhibit enzyme mimicking activity to enhance the rates chemical reactions [19]. These reactions include oxidation or reduction of the substrate and are useful for the development of efficient and reliable colorimetric sensing platforms. Thus, the environmentally relevant reactions are investigated to gain a deeper insight into the operational mechanisms of these innovative technologies, with the focus on the catalytic pathways of nanozymes. The enzyme mimicking functions are classified into substrate oxidation and reduction mechanisms [37]. The activities of oxidase-like and peroxidase-like nanozymes are critical to their functionality. The shape, size, crystal facet, and electronic configuration of these nanoparticles are the determining factors of these catalytic properties. Surface modification techniques are crucial in fine tuning these sensing platforms to enhance their adaptability and effectiveness. Several metal nanoparticles, especially the magnetic iron oxide and zero valent iron nanoparticles, have a bright future in colorimetric sensing applications [19, 20].

4.3 Applications of Iron Nanoparticles

Nanomaterials of bimetallic, trimetallic and multicomponent form, prepared by combination of iron with other noble metals, such as gold, silver and platinum have been developed for various applications. These complexes have not been fully implemented in colorimetric sensing, although they pose advantages including better synthesis economy and flexibility. Bimetallic nanostructures have been applied for the detection of more than 40 analytes, and core–shell nanostructures have been applied for the detection of at least 20 analytes [28, 30]. Many substances can be detected at low concentrations in biological fluids using silver-based sensors. It is highlighted that iron nanostructures can oxidize many analytes and form colorful secondary oxides, and therefore represent a significant opportunity in colorimetric sensors. The incorporation of iron offers additional possible interactions with the products of the detection reaction [1, 9, 11]. Iron oxide nanoparticles are currently one of the most widely used nanoparticles in biomedical fields such as Magnetic Resonance Imaging (MRI), targeted drug delivery, cancer immuno-therapy and hyperthermia mediators. Various biomedical applications are shown in Fig. 4.2. The sensing of 3-nitrotyrosine and lead ions using iron oxides has been demonstrated, and the need for further development in this area will be discussed.

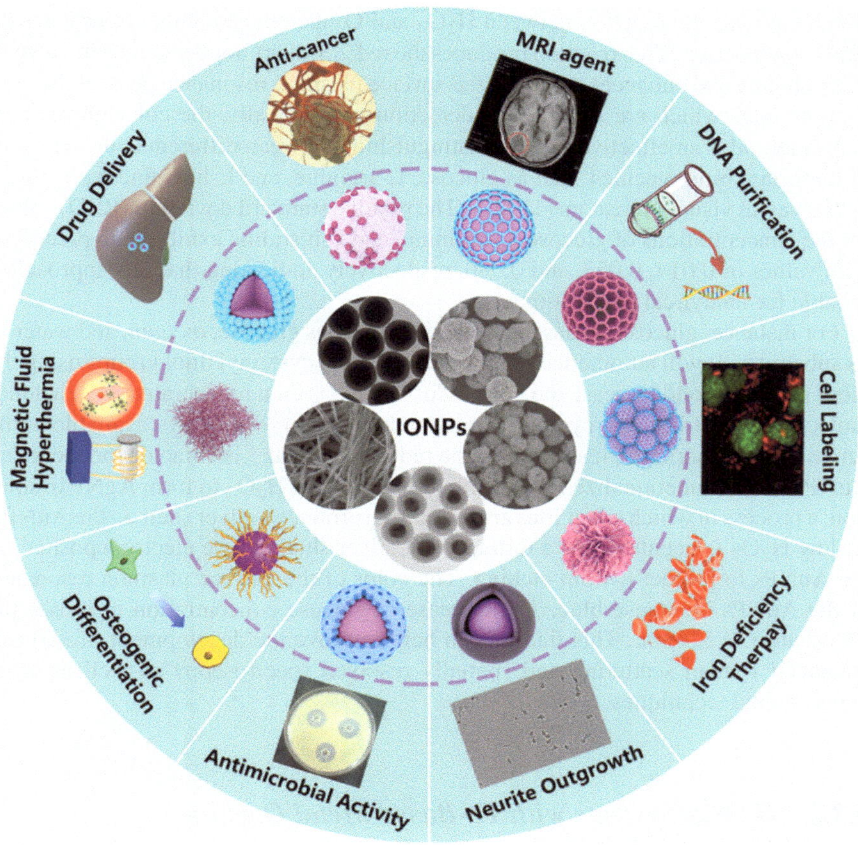

Fig. 4.2 Various biomedical applications of iron oxide nanoparticles. Adapted from [33], Fig. 4.2. Licensed under CC BY 4.0 (https://creativecommons.org/licenses/by/4.0/)

4.3.1 Detection of Hydrogen Peroxide, Glucose, and Other Small Molecules

Numerous metal nanoparticles (MNPs) have been reported to exhibit intrinsic peroxidase-like activity, enabling visible colorimetric changes that are widely utilized for various analytical applications [26, 36]. In our daily lives, there are many small-molecule substrates of important oxidases that play an important role in biochemical processes. However, to detect these small-molecule substrates, it is often necessary to express each of these oxidases separately – a time and resource-consuming process. However, the new approach based on nanomaterials methods enables the simultaneous detection of many of these small-molecule substrates in a more efficient way, these two main methods rely on AuNPs, AgNPs, and even Au–Ag NP for these detections [10, 15]. The most notable are the cavity gold nanoparticles, which catalyze the reaction between H_2O_2 and O-phenylenediamine to produce an oxidation reaction. This reaction produces the red 2,3-diaminophenazine with a vivid color change and enhances the localized surface plasmon resonance peak of the gold nanoparticles which are suitable for detection. Additionally, the colorimetric heat maps method is an effective tool for distinguishing between different concentrations of important biomolecules such as glucose, L-arginine, and L-histidine with simple yet powerful visual indicators [12, 27]. The relative standard deviation (RSD) values for the concentrations of glucose, L-arginine, and L-histidine exhibit high precision, with values of 0.014, 0.053, and 0.080, respectively, indicating that this approach is reliable for analytical applications.

For instance, glucose oxidase is an oxidase that has glucose, oxygen, and water as its substrates. It will automatically decant the protein cysteinyl thiol groups using the these substrates and increase the bound Cu^{2+} which is then regenerated from a donor molecule. It will keep on cycling to ensure that the Cu^{2+} is always available at the active site. Furthermore, it shows that the reduction of Cu^{2+} is cofactor-independent. AuNPS can be functionalized with antibodies to reduce H_2O_2 to form AgNP in solution, a process in which silver ions are reduced to produce a silver shell on the AuNPs. At low H_2O_2 concentrations, a thin layer of silver shell can be electrodeposited on the AuNPs to protect it from etching. The color of the surface plasmon resonance of the AuNPs is intense blue. The increase in glucose concentration increases the absorbance at 526 nm. This finding can help improve the development of surface-enhanced Raman scattering and spatially resolved spectroscopy as well as other experimental techniques.

4.3.2 Hybrid Systems with Fe-Based Metal Organic Frameworks (MOFs)

Structures of Au-Fe-MOF with high yield is prepared and demonstrated to be used in different fields [7]. The Fe-MOFs provide suitable support for gold nanoparticles and satisfied two required criteria for the use of these materials in colorimetric assays, which includes the selective recognition of the analyte and an apparent optical change. It can also provide new opportunities for hyperchromic and plasmonic sensing, thereby indicating the improvement when use in analytical chemistry. The addition of a crimson red Au-Fe-MOF solution to halide ions and inorganic ions (molecules with opposite behavior), for example, can yield a behaviorally unique and eye-catching transition to blue phase. Also, several tests confirmed that the Au-Fe-MOF has excellent performance in identifying Mo^{6+} ions with both high sensitivity and selectivity. This is due to the ability of Fe-MOFs to selectively target the ions of interest through the formation of favorable chemical bonds that enhance their interaction.

It has been reported that iron magnetic MOFs, Au-Fe-MOF is a good candidate for the detection of Mo^{6+} in different practical applications and could also find its place in real-world analytical detection technologies. Numerous researchers have adopted MOFs to construct stable Au nanoparticles for various applications. However, the potential of newly synthesized MOFs has gained importance in recent years, while many features of Fe-MOFs towards specific applications have remained relatively under explored. Although the reliable performance of other MOFs in practical applications is quite well-established, the potential of Fe-MOFs for specific application remains an open question and should be tackled in future work.

Moreover, Au-containing Fe-MOFs were used instead of chloroauric acid as a reducing agent to reduce Au ions. Methods such as UV–Vis spectroscopy demonstrated the efficient detection of I^- with a high detection limit. The synergistic integration of Fe-MOFs, low-resistance nano-Au, and the reversible, high-sensitivity response of Fe-NPs/Tb-MOFs enables practical detection of diverse analytes. This approach supports the development of more generalizable detection methodologies and advances future analytical techniques. Moreover, these findings can go beyond the realm of academia; they can provide guidance for designing new technologies to improve detection capabilities in applications

4.4 Challenges and Future Directions

Ligand exchange plays a critical role in constructing advanced functional platforms. Looking ahead, key priorities include developing portable, robust, user-friendly devices and achieving high-precision, bench-level sensitivity for detecting target species [13, 31]. Response sensitivity is still a crucial limiting factor hindering the practical application of colorimetric sensors. Most response sensors are designed to function at high concentrations of target species, limiting their applicability under varied conditions. Detection systems with high demand and frequent use can be fundamentally important in improving sensitivity and should be prioritized. Detection methods for many medical-associated conditions, such as malignant melanomas, greatly benefit from low detection limits.

In addition, detecting a wider range of bacteria would be greatly advantageous, particularly when utilizing the colorimetric method for rapid and straightforward detection. Secondly, for practical applications, all detection methods should undergo field testing, as such tests are crucial for evaluating performance under real-world environmental conditions. In this and similar situations, implementing measures that enhance portability, usability, and rapid detection will significantly improve the competitiveness of such products, even against established alternatives. Third, the development of multi-utilitarian colorimetric sensing systems may become another pressing and significant aspect area for investigation. Different modes can be efficiently leveraged to dynamically tune detection performance, thereby enabling sensing to adopt to relevant domains and contexts. Overall, this emerging field is advancing rapidly and already offers a set of well-established sensing strategies that are likely to meet current and future needs. As it continues to evolve and adapt to varying needs, this cutting-edge technology holds strong potential for achieving practical, and real-world applications.

4.4.1 Stability, Specificity, and New Avenues for Iron-Based Platforms

Recent studies have shown that FeNPs-based colorimetric sensors may detect tyramine [4, 29]. Tyramine is oxidized, Fe(III) is reduced to Fe(II), and a strong charge transfer complex forms, resulting in a visible color change. This sensing platform which can be fine-tuned through variations in the surface capping of nanoparticles, holds promise for investigating enzyme activities and neurological disorders that are associated with oxidative stress. Iron-based sensing platforms are environmentally friendly due to iron's sustainability and abundance. Nevertheless, some challenges remain; for example, the mixtures require biologically relevant reductants to reduce iron rapidly and efficiently. Studies on catalytic iron nanoparticles may help overcome these limitations and enhancing their application in nanosensing. Compared to other nanoparticles, such as AuNPs and AgNPs, FeNPs are not only cheaper but also non-toxic, making them suitable for large scale applications. However, they may lack sensitivity and stability, and hence require further improvement. There are three key issues must be addressed to optimize their use: (1) Stabilizing and protecting nanoparticles for sustained functionality, (2) Enhancing specificity to minimize interference from other compounds or biological matrices, and (3) Reducing toxicity. Iron-based colorimetric platforms with enhanced surface engineering and nanoparticle features hold potential in diagnostics, catalysis, and environmental assessment, while remaining environmentally friendly and economical.

References

1. Anwar, A., Imran, M., Iqbal, H.M.N.: Smart chemistry and applied perceptions of enzyme-coupled nano-engineered assemblies to meet future biocatalytic challenges. Coord. Chem. Rev. (2023). https://doi.org/10.1016/j.ccr.2023.215329
2. Attia, N.F., Abd El-Monaem, E.M., El-Aqapa, H.G., Elashery, S.E., Eltaweil, A.S., El Kady, M., El-Seedi, H.R.: Iron oxide nanoparticles and their pharmaceutical applications. Appl. Surf. Sci. Adv. **11**, 100284 (2022). https://doi.org/10.1016/j.apsadv.2022.100284
3. Baig, N., Kammakakam, I., Falath, W.: Nanomaterials: a review of synthesis methods, properties, recent progress, and challenges. Mater. Adv. (2021). https://doi.org/10.1039/d0ma00807a
4. Balkourani, G., Brouzgou, A., Tsiakaras, P.: A review on recent advancements in electrochemical detection of dopamine using carbonaceous nanomaterials. Carbon (2023). https://doi.org/10.1016/j.carbon.2023.118281
5. Bilal, M., Khaliq, N., Ashraf, M., Hussain, N., Baqar, Z., Zdarta, J., Iqbal, H.M.: Enzyme mimic nanomaterials as nanozymes with catalytic attributes. Colloids Surf. B **221**, 112950 (2023). https://doi.org/10.1016/j.colsurfb.2022.112950

6. Chong, Y., Liu, Q., Ge, C.: Advances in oxidase-mimicking nanozymes: classification, activity regulation and biomedical applications. Nano Today (2021). https://doi.org/10.1016/j.nantod.2021.101076
7. Fu, D.Y., Liu, X., Zheng, X., Zhou, M., Wang, W., Su, G., Xie, Z.: Polymer-metal-organic framework hybrids for bioimaging and cancer therapy. Coord. Chem. Rev. **456**, 214393 (2022). https://doi.org/10.1016/j.ccr.2021.214393
8. Gebre, S.H.: Bio-inspired synthesis of metal and metal oxide nanoparticles: the key role of phytochemicals. J. Cluster Sci. (2023). https://doi.org/10.1007/s10876-022-02276-9
9. Guillemard, L., Kaplaneris, N., Ackermann, L., Johansson, M.J.: Late-stage C-H functionalization offers new opportunities in drug discovery. Nat. Rev. Chem. **5**(8), 522–545 (2021). https://doi.org/10.1038/s41570-021-00300-6
10. Hang, Y., Wang, A., Wu, N.: Plasmonic silver and gold nanoparticles: shape-and structure-modulated plasmonic functionality for point-of-caring sensing, bio-imaging and medical therapy. Chem. Soc. Rev. (2024). https://doi.org/10.1039/D3CS00793F
11. Hong, S.Y., Hwang, Y., Lee, M., Chang, S.: Mechanism-guided development of transition-metal-catalyzed C–N bond-forming reactions using dioxazolones as the versatile amidating source. Acc. Chem. Res. **54**(11), 2683–2700 (2021). https://doi.org/10.1021/acs.accounts.1c00198
12. Jeon, H.J., Kim, H.S., Chung, E., Lee, D.Y.: Nanozyme-based colorimetric biosensor with a systemic quantification algorithm for noninvasive glucose monitoring. Theranostics (2022). https://doi.org/10.7150/thno.72152
13. Kang, T., Lu, J., Yu, T., Long, Y., Liu, G.: Advances in nucleic acid amplification techniques (NAATs): COVID-19 point-of-care diagnostics as an example. Biosens. Bioelectron. (2022). https://doi.org/10.1016/j.bios.2022.114109
14. Krishnan, S.: Colorimetric visual sensors for point-of-needs testing. Sens. Actuators Rep. (2022). https://doi.org/10.1016/j.snr.2022.100078
15. Kumalasari, M.R., Alfanaar, R., Andreani, A.S.: Gold nanoparticles (AuNPs): a versatile material for biosensor application. Talanta Open (2024). https://doi.org/10.1039/D3CS00793F
16. Kumar, N., Salehiyan, R., Chauke, V., Botlhoko, O.J., Setshedi, K., Scriba, M., Ray, S.S.: Top-down synthesis of graphene: a comprehensive review. FlatChem **27**, 100224 (2021). https://doi.org/10.1016/j.flatc.2021.100224
17. Leonel, A.G., Mansur, A.A.P., Mansur, H.S.: Advanced functional nanostructures based on magnetic iron oxide nanomaterials for water remediation: a review. Water Res. (2021). https://doi.org/10.1016/j.watres.2020.116693
18. Li, Z., Hou, J.T., Wang, S., Zhu, L., He, X., Shen, J.: Recent advances of luminescent sensors for iron and copper: platforms, mechanisms, and bio-applications. Coord. Chem. Rev. **469**, 214695 (2022). https://doi.org/10.1016/j.ccr.2022.214695
19. Liu, Q., Zhang, A., Wang, R., Zhang, Q., Cui, D.: A review on metal-and metal oxide-based nanozymes: properties, mechanisms, and applications. Nano-Micro Lett. (2021). https://doi.org/10.1007/s40820-021-00674-8
20. Lopez-Cantu, D.O., González-González, R.B., Melchor-Martínez, E.M., Martínez, S.A.H., Araújo, R.G., Parra-Arroyo, L., Iqbal, H.M.: Enzyme-mimicking capacities of carbon-dots nanozymes: properties, catalytic mechanism, and applications—a review. Int. J. Biol. Macromol. **194**, 676–687 (2022). https://doi.org/10.1016/j.ijbiomac.2021.11.112
21. Loshchinina, E.A., Vetchinkina, E.P., Kupryashina, M.A.: Diversity of biogenic nanoparticles obtained by the fungi-mediated synthesis: a review. Biomimetics (2022). https://doi.org/10.3390/biomimetics8010001
22. Lyu, Z., Ding, S., Du, D., Qiu, K., Liu, J., Hayashi, K., Lin, Y.: Recent advances in biomedical applications of 2D nanomaterials with peroxidase-like properties. Adv. Drug Deliv. Rev. **185**, 114269 (2022). https://doi.org/10.1016/j.addr.2022.114269
23. Mourdikoudis, S., Kostopoulou, A., LaGrow, A.P.: Magnetic nanoparticle composites: synergistic effects and applications. Adv. Sci. **8**(12), 2004951 (2021). https://doi.org/10.1002/advs.202004951

24. Nguyen, M.D., Tran, H.-V., Xu, S., Lee, T.R.: Fe_3O_4 nanoparticles: structures, synthesis, magnetic properties, surface functionalization, and emerging applications. Appl. Sci. **11**(23), 11301 (2021). https://doi.org/10.3390/app112311301
25. Peters, M.D., Godfrey, C., McInerney, P., Khalil, H., Larsen, P., Marnie, C., Munn, Z.: Best practice guidance and reporting items for the development of scoping review protocols. JBI Evid. Synth. **20**(4), 953–968 (2022). https://doi.org/10.11124/JBIES-21-00242
26. Noreldeen, H.A., Yang, L., Guo, X.Y., He, S.B., Peng, H.P., Deng, H.H., Chen, W.: A peroxidase-like activity-based colorimetric sensor array of noble metal nanozymes to discriminate heavy metal ions. Analyst **147**(1), 101–108 (2022). https://doi.org/10.1039/D1AN01895G
27. Orouji, A., Ghasemi, F., Bigdeli, A., Hormozi-Nezhad, M.R.: Providing multicolor plasmonic patterns with Au@ Ag core–shell nanostructures for visual discrimination of biogenic amines. ACS Appl. Mater. Interfaces **13**(17), 20865–20874 (2021). https://doi.org/10.1021/acsami.1c03183
28. Rajeev, R., Datta, R., Varghese, A., Sudhakar, Y.N., George, L.: Recent advances in bimetallic based nanostructures: synthesis and electrochemical sensing applications. Microchem. J. **163**, 105910 (2021). https://doi.org/10.1016/j.microc.2020.105910
29. Sargazi, S., Fatima, I., Kiani, M.H., Mohammadzadeh, V., Arshad, R., Bilal, M., Behzadmehr, R.: Fluorescent-based nanosensors for selective detection of a wide range of biological macromolecules: a comprehensive review. Int. J. Biol. Macromol. **206**, 115–147 (2022). https://doi.org/10.1016/j.ijbiomac.2022.02.137
30. Stephanie, R., Kim, M.W., Kim, S.H., Kim, J.K., Park, C.Y., Park, T.J.: Recent advances of bimetallic nanomaterials and its nanocomposites for biosensing applications. TrAC, Trends Anal. Chem. **135**, 116159 (2021). https://doi.org/10.1016/j.trac.2020.116159
31. Tobias, D.K., Merino, J., Ahmad, A., Aiken, C., Benham, J.L., Bodhini, D., Nolan, J.J.: Second international consensus report on gaps and opportunities for the clinical translation of precision diabetes medicine. Nat. Med. **29**(10), 2438–2457 (2023). https://doi.org/10.1038/s41591-023-02502-5
32. Qian, S., Cui, Y., Cai, Z., Li, L.: Applications of smartphone-based colorimetric biosensors. Biosens. Bioelectron. X (2022). https://doi.org/10.1016/j.biosx.2022.100173
33. Xu, G., Song, P., Xia, L.: Detection of heavy metal ions based on surface-enhanced Raman scattering spectroscopy. Nanophotonics **10**(15), 3895–3914 (2021). https://doi.org/10.1515/nanoph-2021-0363
34. Yang, F., Du, M., Yin, K., Qiu, Z., Zhao, J., Liu, C., Pang, H.: Applications of metal-organic frameworks in water treatment: a review. Small **18**(11), 2105715 (2022). https://doi.org/10.1002/smll.202105715
35. Yang, J., Wang, X., Sun, Y., Chen, B., Hu, F., Guo, C., Yang, T.: Recent advances in colorimetric sensors based on gold nanoparticles for pathogen detection. Biosensors (2022). https://doi.org/10.3390/bios13010029
36. Ye, M.L., Zhu, Y., Lu, Y., Gan, L., Zhang, Y., Zhao, Y.G.: Magnetic nanomaterials with unique nanozymes-like characteristics for colorimetric sensors: a review. Talanta (2021). https://doi.org/10.1016/j.talanta.2021.122299
37. Zhang, X., Li, G., Chen, G., Wu, D., Wu, Y., James, T.D.: Enzyme mimics for engineered biomimetic cascade nanoreactors: mechanism, applications, and prospects. Adv. Func. Mater. **31**(50), 2106139 (2021). https://doi.org/10.1002/adfm.202106139

Chapter 5
Silver, Platinum, and Palladium Nanoparticles in Colorimetric Systems

Abstract Metal-based nanoparticles such as silver (AgNPs), platinum (PtNPs), and palladium (PdNPs) have emerged as a key players in the development of colorimetric sensing technologies due to their unique plasmonic properties, significant catalytic activity, and adaptability for detecting a wide range of analytes. It also covers their size, shape, and surface properties, that determine their optical properties and sensing efficacy in colorimetric detection. Owing to their pronounced surface plasmon resonance, AgNPs have been extensively utilized for the detection of biomolecules, ions, and small molecules, while PtNPs and PdNPs have emerged as promising materials for sensing applications, especially in the fields of environmental monitoring and biomedical diagnostics. Because nanoparticle-based sensors exhibit high sensitivity, reusability, and stability, this chapter discusses the aggregation and signal amplification mechanisms in NPs-based sensors. We also review the conjugation of these nanoparticles to other biosensor platforms (e.g., electrochemical, field-effect transistor (FET), etc.) that significantly improve detection capabilities. Future efforts will focus on improving the stability, specificity and scalability of nanoparticles for real-world implementation.

Due to localized surface plasmon coupling, metal nanoparticles are very effective for colorimetric sensing. The challenges related to the development of high-quality nanoparticles, the fundamental nature of plasmon enhancement, surfactants coating and capping of, and the solution-phase characteristics of chemical reactions were examined. Potential applications include soil testing, environmental monitoring of VOCs, red tide algae, detection of animal venoms, and identification of biological compounds such as nucleotides, DNA, RNA, proteins, bacteria, and viruses [10, 25, 26]. A discussion on the dynamic range of detection, lower detection limit (LOD), reusability and stability of the method was also included. This chapter therefore examined the current advantages and disadvantages regarding mechanism and application and outlined possible future directions before concluding.

The metal nanoparticle catalysts examined in this chapter include the metals silver, gold, copper, platinum and palladium. The principles, case studies, and influencing

A. C. Alguno et al., *Metal Nanoparticles-Based Nanoplatforms for Colorimetric Sensing*, Engineering Materials, https://doi.org/10.1007/978-981-96-6771-0_5

factors were discussed in detail. The sensing performances of the reported colorimetric sensors were compared for the efficiency and the issues that may limit their practical usability were also addressed. This findings support applications beyond colorimetry, especially in conjunction with other analytical instrumental methods, as sensor elements such as Au, Ag, Pd, and Pt can also contribute to signal generation. Although, it is seldom fully substantiated, the signal detection processes offer opportunities for exploration and continue to attract scientific interests. While many studies focus on silver and copper nanosensors, we also will discuss the potential use of other nanoparticles such as palladium and platinum as future sensing nanoplatforms.

5.1 Silver Nanoparticles (AgNPs)

The size and shape of AgNPs presented in Fig. 5.1 have been investigated using transmission electron microscopy (TEM) and scanning electron microscope (SEM) images, and revealing that the nanoparticles exhibit spherical, rod-like and cubic morphologies. These images highlight the nanoscale features and surface characteristics of the particles, which are relevant to their potential applications. The unique optical properties of AgNPs make them an excellent candidates for colorimetric sensors, they have been exploited for the detection of a broad spectrum of targets such as DNA, multiple antigens, various proteins, small biomolecules and essential ions [2, 11]. The surface plasmon resonance of silver NPs are identified by strong absorptive surface plasmon spectrum, sensitive to the dimensions of the nanostructure [5, 17]. This contributes to broad UV–Vis absorbance spectra that greatly enhance their potential for accurate photometric monitoring applications. In addition, they have strong fluorescence quenching activity which make them recommendable for DNA and protein detection in various assays. Regarding ionic sensitivities, the intrinsic negative charge nanoparticle must also be considered. This property enables their analogs to be used for detecting various targets such as RNA or microRNA, circulating tumor cells, specific nucleotide sequences, bovine DNA, and DNA-Auramine-labeled *Mycobacterium tuberculosis* complex. These applications cover a wide variety of uses, including the hematology laboratory typing methods for highly sensitive detection of DNA; research of complex gene regulations; sensing and bioimaging of mRNA expression levels; detection of telomerase activity, such as single-nucleotide mismatch and single telomerases in various biological samples.

5.1.1 Properties and Role in Aggregation-Based Sensing

Among many methods, metal nanoparticles (NPs)-based aggregation is the most common due to its high efficiency and reliability [4]. We shall also dwell on the size, shape, and concentration dependent properties of the metal NPs, especially that of gold, silver, copper, and platinum NPs which make them all ideal candidates

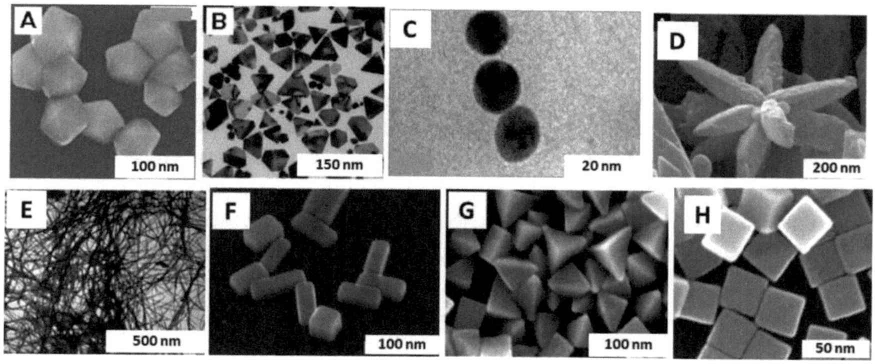

Fig. 5.1 Multiple AgNP forms are shown in transmission electron microscopy (TEM) and scanning electron microscopy (SEM) images. Decahedrons (**a**) and prisms (**b**). The photos also include spheres (**c**), flowers (**d**), nanowires (**e**), nano-bars (**f**), pyramids (**g**), and nano-cubes (**h**). Adapted from [1], Fig. 5.1. Licensed under CC BY 4.0 (https://creativecommons.org/licenses/by/4.0/)

for aggregation-based sensing applications. Moreover, we will discuss the additional functions of these metal NPs as colorimetric sensing agents, acting as either indirect sensing reagents (ISEs), direct sensing reagents (DSEs), or templates. We will demonstrate the high affinity of metal NPs towards various analytes, along with their lower detection limits by several orders of magnitude, highlighting their potential real world applications. Furthermore, we will extensively review the numerous advantageous offered by metal NPs-based aggregation toward colorimetric strategies. We will then suggest several unexplored fields within the metal NP-based sensing domain to lead future research to further study this important area.

Metal NPs display unexpected and significantly different properties in comparison to their bulk materials counterparts [13]. These difference arise from quantum effects, surface phenomena, and macroscopic quantum transport effects. In species that cannot be resolved in traditional sensing mode, miniaturization to the nanoscale reveals novel properties absent in the bulk form. These properties include, but are not limited to, vivid coloration, increased catalytic activity, improved electrical conductivity, altered melting points, superparamagnetism, changes in depolarization temperatures, increased stability, and increased melting transverse relaxation rate [9, 15]. All these properties depends on the concentration, size and shape of the nanoparticles. The unique and tunable property of metal NPs allows potential application in different fields like chemistry, physics, materials science, biology, engineering, and energy. Metal NPs hold great potential as one of the vibrant materials in modern science with broad industrial relevance.

5.1.2 Applications in Nucleic Acid and Protein Detection

The detection of nucleic acids and proteins is of great importance in biological systems and atmosphere, with relevance in the diagnosis and treatment, virus detection, and food safety. Several MNPs-based nanoplatforms have been developed for the detection of nucleic acids and proteins, some utilizing colorimetric approaches. This includes AuNP-based systems, DNAzyme-MNPs-based platforms, DNA-sequence-specific detection, nanoprobe- and probe-modulated strategies, nanomaterial-enzyme (ENZ)-based method, and other DNA/MNPs-based nanoprobes, all tailored for colorimetric sensing. In addition, the contributions and common challenges of these nanoplatforms have been discussed. It is important to continue developing such MNP-based nanoplatforms with new sensing applications that offer more simplicity, higher sensitivity, and specificity, and improved recovery and treatment efficiency, particularly by reusing the modified MNPs through magnetic field manipulation.

The optical properties of thiol-DNA/MNPs-based nanocomposites can be tuned and functionalized via DNA receptors to serve as colorimetric sensors for protein detection. AuNPs have proven the ability to enable simple, label-free colorimetric detection of proteins through plasmon resonances properties. The DNAzyme/MNPs-based nanosensor builds on the well-established colorimetric detection of hydrogen peroxide in the presence of ABTS (2,2′-azino-bis(3-ethylbenzothiazoline-6-sulfonic acid) as a substrate utilizing the controllable peroxidase-like activities and formal properties of MNPs, application that include protein concentration imaging and cancer cell detection. For instance, nucleic acid biosensor techniques were developed and successfully applied for bacterial detection, based on the unique characteristics of polyaniline-coated MNPs, which address the challenges posed by the whole bacterial cells with heterogeneous structures. Enabled by MNPs-based nanosensors or nanoprobes, these approaches office lower costs, high efficiency and simple operation in comparison with traditional detection methods. Owing to these benefit, they represent strong candidates for colorimetric sensing applications.

5.2 Platinum Nanoparticles (PtNPs)

The preparation of platinum nanoparticles (PtNPs) is less common yet easier than the synthesis of AuNPs or AgNPs. Most commonly, PtNPs are polydispersed and irregular in shape, rather than monodisperse or regularly spaced nanoparticles [7, 18]. Figure 5.2 shows TEM images of various shapes and sizes of PtNPs synthesized by using different plant secondary metabolites: (a) terpenes-mediated nanoparticles, (b) phenolic compounds-mediated particles, and (c) sulfur-containing compound-mediated nanoparticles. Every metabolite imparts a distinct morphology to the PtNPs, that can be derived from plant-based sources and applied in a variety of applications. Although PtNPs have been applied in fuel cell development, their application

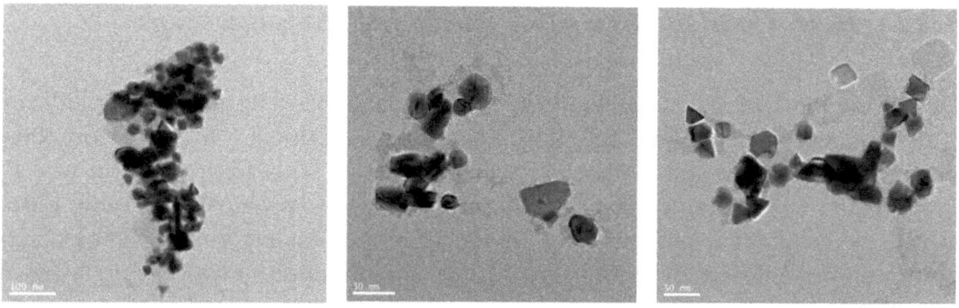

(a) (b) (c)

Fig. 5.2 TEM images show that platinum nanoparticles (PtNPs) can be generated from several plant-derived secondary metabolites, resulting in a variety of forms and sizes. These include **a** terpenes, **b** phenolic compounds, and **c** S-containing compounds. Adapted from [12], Fig. 5.2. Licensed under CC BY 4.0 (https://creativecommons.org/licenses/by/4.0/)

in colorimetric sensing remains underexplored. In addition to flower-shaped PtNPs, core–shell structures of platinum nanoparticles are also being synthesized [16]. These core–shell PtNPs are very promising nanocomposites with unique and valuable properties for colorimetric detection. Further, when combined with transparent substrates colorimetric sensing platforms, these systems can achieve remarkable sensitivity while minimizing the interference commonly encountered in platinum nanoparticles detection and analysis.

Some electrochemical experiments have demonstrated that monolayer PtNPs can be excited under ultraviolet light and efficiently reduce oxygen molecules. This reduction leads to the formation of reactive species such as superoxide anion radicals and hydrogen peroxide, which are extremely relevant in biological systems. Subsequently, PtNPs can absorb visible light (wavelength > 600 nm), generating a singlet excited state through intersystem crossing, thereby functioning as effective photosensitizers with high photodynamic inactivation efficiency against cancer cells. In addition, these nanoparticles exhibit strong interactions with microbial proteins and nanozyme-catalyzed photothermal inactivation under visible light, highlighting their potential use in biological and biomedical applications as targeting agents for harmful cells or pathogens. However, the near-infrared (NIR) light absorption of PtNPs is relatively low, which limits their photothermal efficiency. Consequently, no PtNPs have been reported to efficiently induce surface plasmon resonance. Nonetheless, PtNPs have been developed for numerous analytic applications to address some of the limitations associated with AuNPs or AgNPs in colorimetric detection. Furthermore, existing modification protocols for other metallic NPs must be improved in order to facilitate the effective attachment of PtNPs onto sensor matrices. This aims to provide an overview of the potential of PtNPs as colorimetric sensors across a wide range of applications.

5.2.1 High Catalytic Activities and Signal Amplification

One of the most striking characteristics of metal nanoparticle-based nanoplatforms is their exceptional catalytic activities, which plays a critical role a in colorimetric sensing applications and cannot be overemphasized. This high catalytic activity is attributed to the short distances between surface atom and their atomic cores, as well as the fact that surface atoms are not fully coordinated or enclosed by neighboring atoms unless they establish a defined coordination network. For instance, an ultrafine AuNPs, one or several gold Au atoms in the first layer can form a seven-coordinate network at the center of the nanoparticle [14], a model proposed for their enhanced reactivity. This increased in coordination number is a key factor contributing to the catalytic properties of these nanoparticles. Furthermore, traditional electrochemical techniques, coupled with detailed investigation of Au–S interactions, have demonstrated that the 8-atom Au contact effect is dominates and is the key element in the observed enhancement of catalytic activity.

Beyond the catalytic activities extensively discussed above, many new methods have been reported to induce significant signal amplification using metal NP-based nanoplatforms. Signal amplification is more sophisticated concept, whereby the signal produced from the interaction between analytes and the sensing materials bound to metal nanoparticles is further enhanced, either engineered design of the nanoparticles themselves or through additional chemical or physical process applied in the initial sensing event. In either case, the resulting sensitivity is significantly higher than without amplification. This enhancement is particularly critical for applications requiring precise identification and quantitation in complex matrices.

Although this work has emphasized on the tuning of metal NP based nanoplatforms for colorimetric sensing, it should be noted that these nanomaterials possess additional valuable properties with potential applications in other fields. For instance, metallic nanoparticles have high surface area to volume ratio, making them highly reactive and excellent catalysts in numerous chemical reactions. Moreover, these nanoparticles can be easily conjugated to biomolecules including proteins, DNA, or enzymes, making them useful tools for biosensing applications. Additionally, metal nanoparticles can enhance the performance of fluorescent dyes or biomolecules, which hold significant potential for biological imaging and diagnostics. Thus, this paves the foundation for designing highly sensitive and selective visual-based sensing methods utilizing metal nanoparticle-based nanoplatforms for the analysis of diverse analytes in different sample types.

5.2.2 Detection of Environmental and Toxic Analytes

Singh et al. [24] and Alberti et al. [3] recently reported an important advancement in the development of metal nanoparticle-based colorimetric sensors for effective colorimetric detection of numerous environmental and toxic analytes. These analytes include heavy metal ions, small organic molecules, anions, and well-known environmental toxins and pollutants, long with their degradation products. The effects of heavy metal ions and heavy metal-containing nanomaterials are of immense importance to human health. While the detection capability of recently developed colorimetric sensors have been elaborated for a wide range of environmental and toxic analytes the breath of findings make it challenging to summarize them comprehensively in a single subsection. Hence, this subsection emphasizes on recent discoveries involving existing metal nanoparticle-based colorimetric sensor systems, focusing on emerging environmental contamination and toxic species which significantly influenced by human activities, an area considered as a critical and promising field of research. It is also worth noting that results, apart from metal nanoparticle-based sensors presented in this subsection, many other types of sensors employing different working principles have been developed for general chemosensing applications. Due to their intrinsic properties, metal-based colorimetric sensors offer specific advantages over other types sensors, such as those based on organic or redox species, particularly in terms of their sensitivity toward certain environmental analytes. While we summarize in this subsection, this does not aim to be exhaustive, but it seeks to provide a representative overview that highlights the practical utility and real-world impact of metal nanoparticle-based colorimetric sensors through selected examples.

5.3 Palladium Nanoparticles (PdNPs)

PdNPs were synthesized using plant extracts and characterized by UV–Vis spectroscopy and TEM microscopy. The UV-Vis spectra of PdNPs synthesized using *Evolvulus alsinoides*, *Solanum nigrum* and *Rosmarinus officinalis* are shown in Fig. 5.3a–c. In addition, the unique morphologies and size distributions observed in previous images from this experimental set illustrated in Fig. 5.3d–f. These results demonstrate the potential of green synthesis approaches for producing PdNPs with unique properties and suggest that the properties of the synthesized nanoparticles are influenced by the extraction parameters used to obtain the plant extract.

Another novel platforms for colorimetric sensing involve the use of palladium nanoparticles (PdNPs) through their redispersion in solution [19]. PdNPs are generally stabilized with anionic surfactants. Upon the addition of target reagents, the fluorescence recovery of PdNP-quenched probes may be induced, enabling PdNP-based colorimetric detection. This detection mechanism relies either on the aggregation–redispersion behavior of PdNPs in solutions or on NP-to-alloy transformation, both of which result in a visible color change [20]. With slight modification such sensor can be implemented for different targets. The flexibility of the PdNP platform allows for both fluorescence recovery in DNA-modified PdNPs and the colorimetric detection of small molecules by exploiting electrostatic interactions between the positively charged target–ligand complexes and negatively charged PdNPs.

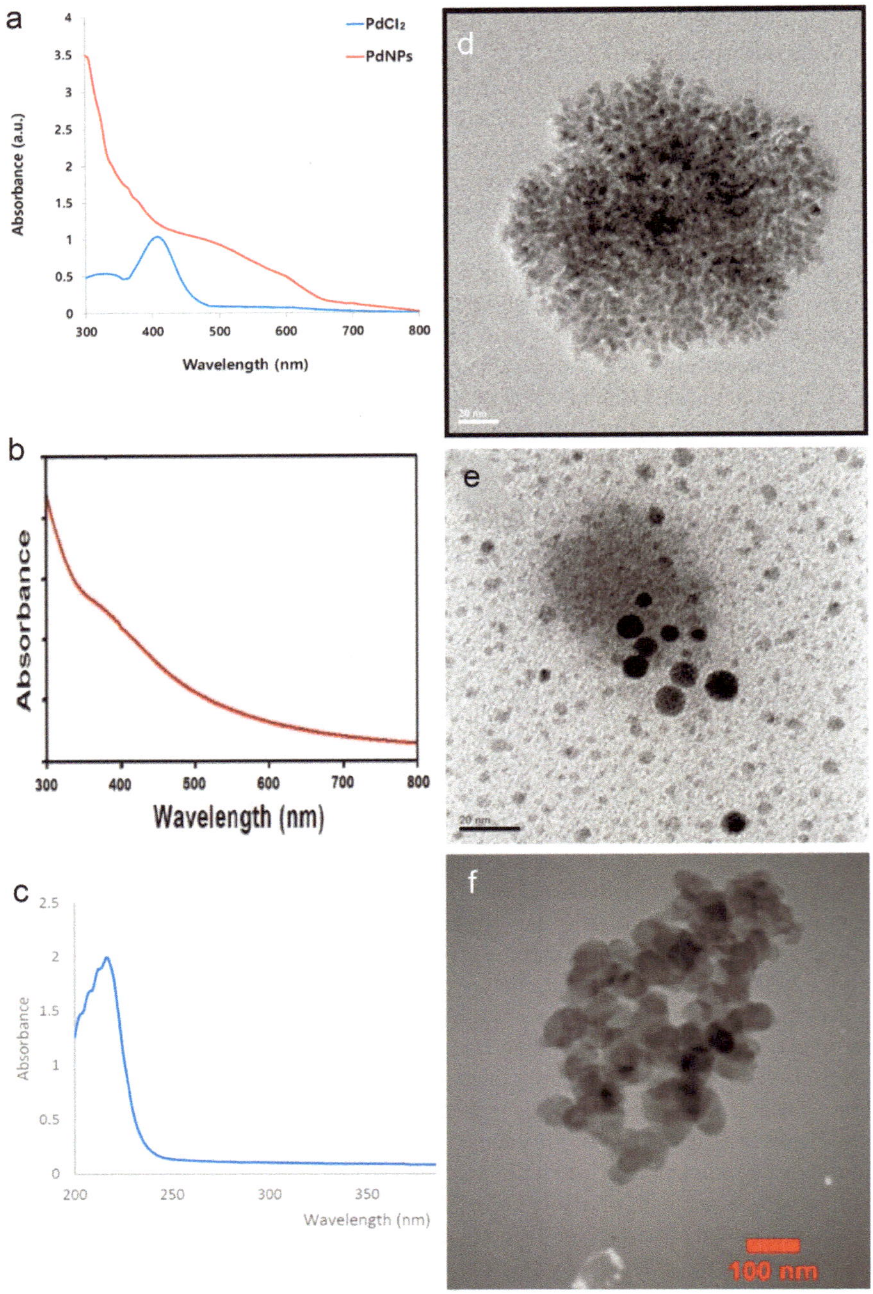

Fig. 5.3 The UV-Vis spectra and TEM pictures of PdNPs that were made using different plant extracts are shown. Spectra (**a–c**) include PdNPs that come from *Evolvulus alsinoides*, *Solanum nigrum*, and *Rosmarinus officinalis*. In the same way, the shape of these nanoparticles from the same plant sources is shown in the TEM photos (**d–f**). Adapted from [8], Fig. 5.3. Licensed under CC BY 4.0 (https://creativecommons.org/licenses/by/4.0/)

PdNP platforms for colorimetric detection, however, are still in their infancy [21]. One important parameter for improving this type of colorimetric sensing is minimizing the aggregation–redispersion behavior, which is a key factor for enhancing the detection sensitivity and stability of PdNPs in the presence of target reagents. Hence, the advantages and limitations of PdNP platforms for colorimetric sensing are evaluated to summarize the latest developments and highlight the weaknesses of this sensing system. Recently developed PdNP-based colorimetric sensors for detecting different environmental targets are discussed, with special attention given to their practical applications, detection methods, sensitivities and potential internal use. Finally, future directions PdNP-based colorimetric sensing are outlined, along with the discussion of the broader limitations of metal NP-based colorimetric sensor in practical applications.

5.3.1 Enzyme Mimicking Properties and Multi-target Detection

The use of metal nanomaterials has enabled the mimicry of natural enzymes to serve as catalysts in the presence of different substrates, and has significantly enhancing their utility for the development of novel colorimetric sensors for the selective and sensitive detection of analytes [28]. For example, several studies have reported effective peroxidase-mimicking activity of reduced three-dimensional graphene oxide-gold nanoparticles (rGO-AuNPs), and a colorimetric method for glucose detection in solution has been developed [27]. The addition of glucose led to the formation of blue products and a significant decrease in absorbance values. This decrease in absorbance was exploited to establish the colorimetric method, which demonstrated a linear detection range from 50.0 μM to 10. 0 mM, making it highly suitable for practical applications.

Moreover, nitrogen-doped porous carbon nanosheets decorated with platinum metal nanoparticles demonstrate excellent enzyme-like properties. Accordingly, a dual-enzyme mimetic nanoplatform was proposed for the simultaneous colorimetric detection of glucose and hydrogen peroxide. The system can reduce oxygen to water (H_2O) and oxidise para-chloroaniline (PCA) to para-aminobenzoic acid (PAC) through the catalytic activity of the nanoparticles, functioning as an enzyme sensor for β-galactosidase activity. The nanoparticles, decorated with numerous BSA and horseradish peroxidase–labeled secondary antibodies (HRP-Ab_2), were arranged in a sandwich configuration, resulting in substantial enhancement of both surface plasmon resonance (SPR) and catalytic efficiency of the nanoparticles to oxidize 3,3,5,5′-tetramethylbenzidine (TMB). The system demonstrated strong detection performance for β-galactosidase activity, with a detection limit of 3.8 nM. Although such studies reveal certain drawbacks of metal nanomaterial-based detection methods, further insights into their enzymatic behavior are essential to drive further advancement in next-generation sensing technologies.

5.3.2 Integration in Advanced Biosensor Designs

A promising new trend in the improvement of sensitivity of sensing systems involves the integration of AuNPs with various biosensors platforms, including field-effect transistor (FET) and electrochemical biosensors. The size and shape AuNPs are easily controllable their surface s are chemically stable and their unique optical properties make them particularly suitable for use in FET biosensors. This integration is particularly useful for signal amplification, direct electron transfer, and replacing conventional charge transfer media to enhance biosensor sensitivity. Because of the ease of controlling their size and stability during the synthesis and surface modification, AuNPs have been employed as sensitive transducers in the development of FET biosensors. Their tunable characteristics allow them to be tailored to specific requirements, thereby enabling the design of new biosensing technologies capable of operating under different conditions and detecting various analytes with high sensitivity and accuracy.

Research studies have also explored the innovative application of metal nanoparticles in the development of electrochemical biosensors for the detection of biologically-relevant compounds including small molecules, DNA hybridization and cell-related processes. In one such experiment, a two-dimensional layer-by-layer assembly technique was employed to construct a mixed azocalix[4]arene/Al-Schiff base adenine derivative, which was deposited in a polyelectrolyte matrix on the glassy carbon electrode. AuNPs were subsequently deposited onto the electrode-posited polyelectrolyte multilayers. Remarkably, these AuNPs enhanced both the electron transfer rate and the interfacial electron transfer kinetics of the biosensor system. The results of this study are quite useful and contribute significantly to advancing the integration of AuNPs into the design of high-performance electrochemical biosensors. Such improvements are critical for improving the sensitivity, efficiency, and overall performance of biosensing systems across a wide range of applications.

5.4 Comparison and Complementary Use of Ag, Pt, and Pd Nanoparticles

To this end, several representative sensing systems based on Ag, Pt, and Pd nanoparticles are detailed and illustrated [17, 22]. This section particularly highlights the unique properties of Ag nanoparticle a presents a wide range of strategies aimed at addressing existing challenges in sensing technology. Additionally, a comprehensive comparison is provided, underscoring the comparative advantages and inherent limitations of these sensing systems, as well as important factors that may significantly affects the reactivity and performance of Ag, Pt, or Pd nanoparticles. Moreover, the use of carbon dots and core–shell nanoparticles (Au@Pt) as precatalysts in

AgNP based systems has led to significant improvements in the colorimetric detection of glucose or hydrogen peroxide (H_2O_2), enhancing both sensitivity and overall analytical performance [6]. The unique catalytic activities of each type of noble metal nanoparticle type are well-organized within respective application domains. Notably, in dose-dependent colorimetric assays, AgNPs exhibited earlier color decay, implying relatively higher sensitivity and response amplitude as a function of reagent concentration, an attribute that supports their suitability for sensing applications. The growing availability of these metal nanoparticles in the colorimetric sensing platforms suggests that the rational activation and integration of multiple noble metal nanoparticles could lead to substantial improvements in the accuracy and reliability of real-world detection systems. A thorough discussion of the catalytic characteristics of metal-based nanoparticles in colorimetric detection may provide valuable insights for refining surface modification strategies and integration methods, ultimately contributing to the development of more effective and efficient sensing devices. If these factors are adequately addressed, significant progress can be made in advancing analytical sensing technologies.

5.4.1 Advantages and Limitations in Various Analytical Contexts

Shruti et al. [23] reported that the enhanced sensitivity is one of the major advantages of nanoparticle-based colorimetric sensing over conventional dye or indicator-based methods. This key property accounts for the high sensitivity of many nanoparticle-based assays. However, the use of size-dependent optical properties of nanoparticles remains limited in qualitative and semi-quantitative assays. A wide range of colorimetric techniques are employed in clinical diagnostics and are equally useful in environmental monitoring, food safety surveillance, security, and anti-terrorism applications. These traditional methods are often quantitative or semi-qualitative and are typically designed for visual inspection. Conversely, nanoparticle-based sensing platforms are generally used when more analytical precision is needed, as a monitoring color changes in nanoparticle suspensions usually necessitates a calorimetric device or a cuvette-based measurement. In the absence of sufficient sensitivity, the distribution of nanoparticles in the medium does not significantly influence their optical properties or the resulting signal, especially in samples with higher Raman enhancement. Consequently, the qualitative analysis of mixtures containing different particle types or aged, inhomogeneous particles can be complex. Nanoparticles also exhibit a significant increase in photo-enhanced fluorescence (PEF) and exhibit aggregation-quenching inhibition effects. Solutions containing specific nanoparticle species produce discernible color changes upon interaction with target analytes, indicating that such effects can be observed without the need for individual species separation. Nevertheless, the nanoparticles must be finely dispersed for these effects to become apparent and measurable.

References

1. Abbas, R., Luo, J., Qi, X., Naz, A., Khan, I.A., Liu, H., Yu, S., Wei, J.: Silver nanoparticles: synthesis, structure, properties and applications. Nanomaterials **14**(17), 1425 (2024). https://doi.org/10.3390/nano14171425
2. Ahmed, A., Usman, M., Ji, Z., Rafiq, M., Yu, B., Shen, Y., Cong, H.: Nature-inspired biogenic synthesis of silver nanoparticles for antibacterial applications. Mater. Today Chem. **27**, 101339 (2023). https://doi.org/10.1016/j.mtchem.2022.101339
3. Alberti, G., Zanoni, C., Magnaghi, L.R., Biesuz, R.: Gold and silver nanoparticle-based colorimetric sensors: new trends and applications. Chemosensors (2021). https://doi.org/10.3390/chemosensors9110305
4. Ali, S., Chen, X., Shi, W., Huang, G., Yuan, L.M., Meng, L., Chen, X.: Recent advances in silver and gold nanoparticles-based colorimetric sensors for heavy metal ions detection: a review. Crit. Rev. Anal. Chem. **53**(3), 718–750 (2023). https://doi.org/10.1080/10408347.2021.1973886
5. Bélteky, P., Rónavári, A., Zakupszky, D., Boka, E., Igaz, N., Szerencsés, B., Kónya, Z.: Are smaller nanoparticles always better? Understanding the biological effect of size-dependent silver nanoparticle aggregation under biorelevant conditions. Int. J. Nanomed. 3021–3040 (2021). https://doi.org/10.2147/IJN.S304138
6. Chatterjee, S., Lou, X.Y., Liang, F., Yang, Y.W.: Surface-functionalized gold and silver nanoparticles for colorimetric and fluorescent sensing of metal ions and biomolecules. Coord. Chem. Rev.. Chem. Rev. (2022). https://doi.org/10.1016/j.ccr.2022.214461
7. Dawadi, S., Katuwal, S., Gupta, A., Lamichhane, U., Thapa, R., Jaisi, S., Parajuli, N.: Current research on silver nanoparticles: synthesis, characterization, and applications. J. Nanomater.Nanomater. **2021**(1), 6687290 (2021). https://doi.org/10.1155/2021/6687290
8. Fahmy, S.A., Preis, E., Bakowsky, U., Azzazy, H.M.E.-S.: Palladium nanoparticles fabricated by green chemistry: promising chemotherapeutic, antioxidant and antimicrobial agents. Materials **13**(17), 3661 (2020). https://doi.org/10.3390/ma13173661
9. Fritea, L., Banica, F., Costea, T.O., Moldovan, L., Dobjanschi, L., Muresan, M., Cavalu, S.: Metal nanoparticles and carbon-based nanomaterials for improved performances of electrochemical (bio) sensors with biomedical applications. Materials **14**(21), 6319 (2021). https://doi.org/10.3390/ma14216319
10. Gavrilaş, S., Ursachi, C.Ş, Perţa-Crişan, S., Munteanu, F.D.: Recent trends in biosensors for environmental quality monitoring. Sensors **22**(4), 1513 (2022). https://doi.org/10.3390/s22041513
11. Giri, A.K., Jena, B., Biswal, B., Pradhan, A. K., Arakha, M., Acharya, S., Acharya, L.: Green synthesis and characterization of silver nanoparticles using Eugenia roxburghii DC. Extract and activity against biofilm-producing bacteria. Sci. Rep. **12**(1), 8383 (2022). https://doi.org/10.1038/s41598-022-12484-y
12. Jeyaraj, M., Gurunathan, S., Qasim, M., Kang, M.-H., Kim, J.-H.: A comprehensive review on the synthesis, characterization, and biomedical application of platinum nanoparticles. Nanomaterials **9**(12), 1719 (2019). https://doi.org/10.3390/nano9121719
13. Khan, M., Khan, M.S.A., Borah, K.K., Goswami, Y., Hakeem, K.R., Chakrabartty, I.: The potential exposure and hazards of metal-based nanoparticles on plants and environment, with special emphasis on ZnO NPs, TiO_2 NPs, and AgNPs: a review. Environ. Adv. **6**, 100128 (2021). https://doi.org/10.1016/j.envadv.2021.100128
14. Li, Y., An, J.D., Wang, T.T., Shi, Y.F., Huo, J.Z., Wu, X.X., Ding, B.: An ultra-stable Cadmium (II) coordination framework constructed from the new bi-functional ligand and application as fluorescent probe for acetylacetone and antibiotics. Dyes Pigm.Pigm. **186**, 109039 (2021). https://doi.org/10.1016/j.dyepig.2020.109039
15. Li, X., Yue, R., Guan, G., Zhang, C., Zhou, Y., Song, G.: Recent development of pH-responsive theranostic nanoplatforms for magnetic resonance imaging-guided cancer therapy. Exploration (2023). https://doi.org/10.1002/EXP.20220002

16. Lin, X., Yu, W., Tong, X., Li, C., Duan, N., Wang, Z., Wu, S.: Application of nanomaterials for coping with mycotoxin contamination in food safety: from detection to control. Crit. Rev. Anal. Chem. **54**(2), 355–388 (2024). https://doi.org/10.1080/10408347.2022.2076063
17. Montes-García, V., Squillaci, M.A., Diez-Castellnou, M., Ong, Q.K., Stellacci, F., Samori, P.: Chemical sensing with Au and Ag nanoparticles. Chem. Soc. Rev. **50**(2), 1269–1304 (2021). https://doi.org/10.1039/D0CS01112F
18. Mourdikoudis, S., Menelaou, M., Fiuza-Maneiro, N., Zheng, G., Wei, S., Pérez-Juste, J., Sofer, Z.: Oleic acid/oleylamine ligand pair: a versatile combination in the synthesis of colloidal nanoparticles. Nanoscale Horiz. **7**(9), 941–1015 (2022). https://doi.org/10.1039/D2NH00111J
19. Nilghaz, A., Mousavi, S.M., Tian, J., Cao, R., Guijt, R.M., Wang, X.: Noble-metal nanoparticle-based colorimetric diagnostic assays for point-of-need applications. ACS Appl. Nano Mater. **4**(12), 12808–12824 (2021). https://doi.org/10.1021/acsanm.1c01545
20. Rao, H., Xue, X., Luo, M., Liu, H., Xue, Z.: Recent advances in the development of colorimetric analysis and testing based on aggregation-induced nanozymes. Chin. Chem. Lett.. Chem. Lett. (2021). https://doi.org/10.1021/acsanm.1c01545
21. Scott, N.W., Ford, M.J., Jeddi, N., Eyles, A., Simon, L., Whitwood, A.C., Fairlamb, I.J.: A dichotomy in cross-coupling site selectivity in a dihalogenated heteroarene: influence of mononuclear Pd, Pd clusters, and Pd Nanoparticles—the case for exploiting Pd catalyst speciation. J. Am. Chem. Soc. **143**(25), 9682–9693 (2021). https://doi.org/10.1021/jacs.1c05294
22. She, M., Wang, Z., Chen, J., Li, Q., Liu, P., Chen, F., Li, J.: Design strategy and recent progress of fluorescent probe for noble metal ions (Ag, Au, Pd, and Pt). Coord. Chem. Rev.. Chem. Rev. **432**, 213712 (2021). https://doi.org/10.1016/j.ccr.2020.213712
23. Shruti, A., Bage, N., Kar, P.: Nanomaterials based sensors for analysis of food safety. Food Chem. (2024). https://doi.org/10.1016/j.foodchem.2023.137284
24. Singh, H., Bamrah, A., Bhardwaj, S.K., Deep, A., Khatri, M., Brown, R.J., Kim, K.H.: Recent advances in the application of noble metal nanoparticles in colorimetric sensors for lead ions. Environ. Sci. Nano **8**(4), 863–889 (2021). https://doi.org/10.1039/D0EN00963F
25. Sohrabi, H., Hemmati, A., Majidi, M.R., Eyvazi, S., Jahanban-Esfahlan, A., Baradaran, B., de la Guardia, M.: Recent advances on portable sensing and biosensing assays applied for detection of main chemical and biological pollutant agents in water samples: a critical review. TrAC, Trends Anal. Chem. **143**, 116344 (2021). https://doi.org/10.1016/j.trac.2021.116344
26. Thakur, A., Kumar, A.: Recent advances on rapid detection and remediation of environmental pollutants utilizing nanomaterials-based (bio) sensors. Sci. Total. Environ. (2022). https://doi.org/10.1016/j.scitotenv.2022.155219
27. Yang, Y., Li, W., Liu, J.: Review of recent progress on DNA-based biosensors for Pb^{2+} detection. Anal. Chim. ActaChim. Acta (2021). https://doi.org/10.1016/j.aca.2020.12.056
28. Ye, M.L., Zhu, Y., Lu, Y., Gan, L., Zhang, Y., Zhao, Y.G.: Magnetic nanomaterials with unique nanozymes-like characteristics for colorimetric sensors: a review. Talanta (2021). https://doi.org/10.1016/j.talanta.2021.122299

Chapter 6
Hybrid and Composite Metal Nanoparticles for Colorimetric Sensing

Abstract Both hybrid and composite metal nanoparticles (NPs) are considered promising platforms for colorimetric sensing due to their notable sensitivity, stability and functionality. In this chapter, we focus on the hybridization of metal NPs, including gold (Au) and silver (Ag), with diverse organic and inorganic materials to design hybrid nanostructures that enhance detection performance. By exploiting the synergistic plasmonic effects of core–shell and multi-metallic architectures, this approach allows the precise and selective detection of biogenic materials, organic compounds, and hazardous elements. Hybrid systems including Au@Fe_3O_4 and Au@Pt nanoparticles, along with their potential applications in diagnostics, environmental monitoring, and biotechnology are examined in this chapter. The exploration of such systems may significantly impact multiplexed colorimetric assays, real-time detection, and high-throughput screening, all of which benefit from the unique properties of composite nanostructures. This chapter discusses recent advances, existing challenges, and future prospects in hybrid nanoparticle-based sensing, emphasizing their potential to revolutionize diagnostics, biosensing technologies, and environmental safety monitoring.

6.1 Overview of Hybrid and Composite Nanostructures

Due to their unique advantages such as enhanced sensitivity, signal amplification, and improved stability, hybrid nanostructures composed of at least two materials with distinct properties have been widely utilized [20, 32, 33]. We also highlight nanoparticle-based hybrids and composites, emphasizing their diverse combinations in terms of size, shape, and spatial arrangement [22]. There is a great deal of interest in incorporating these nanoparticles with semiconductor nanoparticles, carbonaceous nanostructures, and quantum dots to further enhance functionality. The integration of additional functional nanomaterials broadens the scope of potential applications of these systems. The structural properties and electrons interactions within these hybrid structures continue to drive interest in exploring metal nanoparticle-based platforms [14]. Some possible applications of these innovations

A. C. Alguno et al., *Metal Nanoparticles-Based Nanoplatforms for Colorimetric Sensing*, Engineering Materials, https://doi.org/10.1007/978-981-96-6771-0_6

span basic scientific research, nanotechnology, and other applications. However, the full development and application of this technology depend on achieving a detailed understanding of the underlying nanostructures.

6.1.1 Combining Multiple Metals and Functional Components

Gold (Au) and silver (Ag) are the traditional noble metals used to improve sensing sensitivity in plasmonic-based sensors, although researchers are now exploring alternatives beyond these conventional materials [13, 26]. This new direction, which involves the use of a broader metals, represents a paradigm shift, especially in the context of mixed metals, which are rarely employed in nanoparticle array-based colorimetric sensors. Future studies on arrays of multi-metal nanoparticles may reveal cooperative plasmonic properties that could transform certain sensing technologies. Additionally, the self-assembly of heterometallic nanoparticles provides new possibilities for constructing multisensor systems composed of different metals that exhibit enhanced sensitivity in the near-infrared region [12, 29]. In real world applications, these sensors are designed to be highly selective and capable to differentiate between individual optical signals. Overall, these sensor advancements with technological innovations have the potential to stimulate progress in various scientific and industrial fields and contribute meaningfully to the development of analytical and diagnostic platforms.

6.1.2 Design Strategies for Synergistic Enhancement

The physicochemical properties of both types of nanoparticles should be properly regulated to maximize synergism-driven catalytic functions, as their modulation is a well-known strategy to significantly enhance catalytic activity [18]. Copper-based particles with a large surface area and controlled electronic state exhibit strong peroxidase-like activity, whereas smaller amounts of gold nanoparticles (AuNP)-based possess even higher enzyme-like activity [31, 34]. As a result, ultrathin Au-cored copper-based nanoshells demonstrate excellent peroxidase mimicking activity while monodisperse AuNPs show notable catalase- mimicking properties. Currently, advancements have been made in dual and multi nanoparticle systems particularly involving to iron-based and palladium-based nanoparticles, which cooperate in a synergistic manner of optimization of performance. This leads to unique form of synergistic signal amplification that enable precise and reliable readouts across different systems. This effect is attributed to the increase in colorimetric signals intensity when the gold core size is equal to or greater than 4.0 nm, having a peak at 610 nm. This indicate that nanoparticle size directly influences the colorimetric response and consequently, the accuracy of the final analytical results.

6.2 Hybrid Metal Nanoparticles for Colorimetric Platforms

Advances in nanochemistry have spurred the development of new metal nanoparticles for colorimetric sensing, particularly alloy and hybrid nanoparticles [5, 6]. Among these, hybrid nanoparticles composed of two or more metallic components are notable for their remarkable physicochemical features [11]. Their structural diversity enhances sensitivity and selectivity for colorimetric detection, expanding possibilities in plasmonic sensing, imaging, and biomedical applications. Their excellent responsiveness to environmental changes also makes them promising candidate for real-time, label-free biochemical detection and environmental monitoring. Hybrid metal nanomaterials offer superior analytical performance compared to traditional sensor by leveraging their tunable optical properties and synergistic effects. Recent advances in this field has open a wide range of potential applications, including diagnostics, pollution control, and advanced biosensing technologies. These developments underscore their critical role in advancing accurate, low-cost, and non-invasive detection methods.

6.2.1 *Au@Fe_3O_4, Au@Pt, and DNA-Templated Ag/Pt Nanoclusters*

The primary advantage lies in the highly selective and accurate nature of biomolecular recognition compared to traditional colorimetric methods [14], which is essential for the early detection and treatment of cancer and infectious diseases. Labor and testing costs are reduced when using effective labels such as metal nanoparticles are used to achieve lower detection limits. However, the detection of low-abundance targets in complex biological samples remains a significant challenge, requiring the development of fast, quantitative, and label-free methods [19]. The highlights these developments over the past ten to fifteen years, focusing mainly on gold-based detection systems, where biomolecular recognition elements are immobilized on the surface of functionalized nanoparticles. Metallic or bimetallic core–shell nanoparticles are generally practical, as they can provide multilevel information regarding target concentrations. Bare magnetic NPs, for example, can be functionalized to detect DNA targets and can be enriched and isolated using external magnetic field, thereby improving detection specificity [27]. Innovations in biomolecular recognition and sensing technologies include the development of new materials such as Au@Fe_3O_4, Au@Pt, and DNA-templated Ag/Pt nanoclusters.

6.3 Applications of Hybrid Nanoparticles

The specific information required for the home and office application of biomolecules such as DNA and proteins for the diagnosis and management of diseases is presented through clinical applications [1, 7]. However, the detection of specific target biomolecules is challenging and typically requires expert handling. The family of silver nanoparticles directed by DNA has evolved from the first DNA-modified silver particles, which exhibited adjustable solubility and visible colors changes depending on the DNA chain length [3, 21], to more feasible approaches for sensing applications using DNA or its composites with silver nanoparticles. However, limited sensitivity and accuracy of protein detection continue to hinder broader usability and translation. A hybrid formulation combining mesoporous silica with silver nanoparticles has been shown to significantly enhance both the sensitivity and accuracy of the detection platform [23].

6.3.1 Multi-analyte Detection and Simultaneous Monitoring

It is necessary to detect several targets in a single assay for clinical diagnosis, environmental monitoring, and food safety applications [25]. Simple visual methods are particularly useful for on–site testing, as conventional sample preparation is often time consuming and costly. Therefore, there has been increasing development of multiplexed colorimetric assays to enhance detection capabilities. These requirements are effectively addressed by nanoparticle based nanoplatforms, which enable high-efficiency multiplexed colorimetric detection through unique color pairs differentiation based on DNA melting behavior during solid-phase interaction and molecular beacon technology.

Several multiplexed sensing approaches for miRNA signatures have been reported, which represent a major advancement in molecular diagnostics [4, 35]. Magnetic nanoparticles (MNPs), in particular, have emerged as a robust platform that significantly improves the colorimetric detection of miRNA-induced events (MIEs). Their surface features allow competitive interactions between multiplexed miRNAs and Raman probes, thereby promoting detection sensitivity and specificity. By leveraging the unique magnetic and optical characteristics of MNPs, this method provides a new paradigm for high-throughput, specific miRNA profiling, with strong predictive potential for early diagnosis, biomarker discovery, and next-generation biosensing technologies.

The glass bottom chip exhibits unique wettability and is capable of supporting dual functions. The hydrophilicity of titanium nanoparticles (TNPs) ensures that the entire surface is uniformly covered by fluid [9]. At the same time, aluminum nanoparticles (AgNPs) induce superhydrophobicity, allowing for highly controlled droplet manipulation. The uniform dispersion of the hydrophobic AgNPs on the glass substrate is crucial for achieving high-efficiency multiplexed detection [15,

16]. This combined hydrophilic and hydrophobic property configuration improves microfluidic performance, facilitating liquid handling and processing required in biosensing, lab-on-a-chip systems, and next generation diagnostic devices. Thus, the accuracy of the fluid-based analytical systems is greatly improved.

6.3.2 Point-of-Care Diagnostics and Environmental Applications

Colorimetric assays for the detection of proteins and nucleic acids are widely used for point-of-care-diagnostics [24]. Numerous commercial systems are available based on the use of colloidal gold probes, including the QuickVue Influenza A and B tests, OraQuick ADVANCE HIV 1/2 Antibody Test, SeraQuick Vitrum and i-STAT cTnI test [8, 32, 33]. These assays have transformed rapid diagnostic testing across the entire healthcare system. However, dual colorimetric systems often require the use of secondary colors for analyte verification in lateral flow devices due to certain limitations of the gold colloidal particles. Detail schemes that integrate secondary color bands with surface-enhanced Raman scattering (SERS) platforms are still required. Other technologies contributing towards the development of point-of-care diagnostics include high-sensitivity nanomaterials [17]. By combining the signal strength of colorimetric assays with the molecular selectivity of the Raman effect, innovative approaches to molecular diagnostics are emerging. Various dual colorimetric SERS platforms have been developed for the detection of diverse biomarkers, including proteins and viruses, displaying great applicability. A proof of concept for a high-performance malaria detection based on this approach has been reported [14] which successfully distinguish amyloid fibrils in diluted blood samples. Once the COVID-19 pandemic has subsided, AuNPs may become the material of choice for enhancing the convenience and effectiveness of diagnostics in environmental and public health monitoring.

6.4 Future Directions for Hybrid Nanomaterials

Due to the vividly perceptible colors produced through the localized surface plasmon resonance (LSPR) band, colorimetric sensors based on AuNPs are becoming more prominent in the scientific community [2]. The broad responsiveness of these sensors enables their applications across diverse systems. However, due to high production costs and certain limitations in the properties of AuNPs, researchers have turned to the development of hybrid nanomaterials that incorporate AuNPs. This approach creates new opportunities and increase the sensitivity of colorimetric sensors. Compared to sensors solely of AuNPs, hybrid systems offer complementary properties by merging

AuNPs with advanced materials, thereby improving overall sensor performance in real-world applications.

The integration of AuNPs with different materials in hybrid models, including cytochrome protein, different classes of nucleic acids, organic molecules, functional polymers, magnetic nanoparticles, and semiconductor nanoparticles is ubiquitous [15, 16, 30]. One of the key applications of this approach is the significant amplification of AuNP sensitivity to a broad range chemical and biological species. These material combinations offer advantageous for developing efficient detoxification strategies for environmental remediation and exploring novel approaches in renewable energy harvesting, with potential to transform multiple fields. Nonetheless, some limitations of hybrid nanomaterials have been identified, and their practical use and application remain challenging. Further facilitation and in-depth research are needed to their development and optimization for applied.

Promising directions in the emerging field of soft hybrid nanomaterials are also being explored, offering potential for future applications and innovative research pathways. These advancements influence science, technology, and environmental sustainability and may inspire continued efforts to enhance the performance of AuNPs-based colorimetric sensors.

6.4.1 Developing Novel Structures with Superior Properties

Because of their outstanding and nearly boundless applications in catalysis, optical biosensing, bioimaging, and countless other fields, the design and synthesis of ultrasmall metal nanoparticles with predetermined structures and characteristics are of significant interest. Consequently, researchers continue to purse advanced strategies to synthesize materials with unique structures and enhanced performance [5, 6]. Among the primary methods for preparing the ultrasmall metal nanoparticles are template synthesis, electrochemical techniques, and seed-mediated growth. The newer concept, known as electric-field or magnetic-field catalysis, involves the application of an external electric or magnetic field to influence not only the intrinsic catalytic activity but also the collision frequency between the substrate and the catalyst. These advanced approaches inherently rely on surface chemistry and the ability to create highly active and organized surfaces. Numerous shapes of ultrasmall metal nanoparticles have been synthesized and applied across diverse domains owing to their unique characteristics. In recent decades, ultrasmall metal nanoparticles with high stability, and ease of recovery have attracted considerable attention from scientists and researchers. Advances in materials science and imaging technologies have enabled the synthesis of several classes of metal nanoparticles with well-defined geometries. In this context, ultrasmall metal nanoparticles can be categorized into the following types: nanoparticles with various igand shells; Ultrasmall metal exhibiting half-metallic properties; Ultrasmall rare earth metal nanoparticles; and Ultrasmall alloyed nanoparticles. These developments will not only broaden the scope of their application but also suggest new direction for research and technical innovation.

6.4.2 *Exploring New Functionalization Techniques*

Furthermore, it should be emphasized that the relative quantitation of surface features, particle size, homogeneity, shape, and the particle size distribution are all properties that can be tailored to suit specific applications [10, 28]. The continual evolution in this field present a significant challenge for researchers who are investigating new functionalization strategies to improve these critical properties. Regarding silicon dioxide, it is more accurate to state that the system of interaction between particles is responsible for the overall material properties, rather than attributing these properties solely to the nanoparticles themselves. The complexity arises from the fact that the properties of these particles are fundamentally linked to their atomic arrangement and geometry. The correlated physical and chemical characteristics that emerge from this understanding have become significant research topics across a broad range of applications, with increasing interest and investment in recent years. However, the range of the methods available for functionalizing such labels remains limited in comparison to the number of nanoparticle-based labels currently in use. This discrepancy suggests that some aspects of selective recognition remain underexplored in many nanoparticle systems. Motivated by visions of the future, researchers continue to seek ways to integrate selective molecular recognition functionalities into the magnetic, electronic, and catalytic properties of various nanoparticles classes.

References

1. Ahmad, A., Imran, M., Ahsan, H.: Biomarkers as biomedical bioindicators: approaches and techniques for the detection, analysis, and validation of novel biomarkers of diseases. Pharmaceutics (2023). https://doi.org/10.3390/pharmaceutics15061630
2. Alberti, G., Zanoni, C., Magnaghi, L.R., Biesuz, R.: Gold and silver nanoparticle-based colorimetric sensors: new trends and applications. Chemosensors (2021). https://doi.org/10.3390/chemosensors9110305
3. Biba, R., Košpić, K., Komazec, B., Markulin, D., Cvjetko, P., Pavoković, D., Balen, B.: Surface coating-modulated phytotoxic responses of silver nanoparticles in plants and freshwater green algae. Nanomaterials **12**(1), 24 (2021). https://doi.org/10.3390/nano12010024
4. Cai, S., Pataillot-Meakin, T., Shibakawa, A., Ren, R., Bevan, C.L., Ladame, S., Edel, J.B.: Single-molecule amplification-free multiplexed detection of circulating microRNA cancer biomarkers from serum. Nat. Commun.Commun. **12**(1), 3515 (2021). https://doi.org/10.1038/s41467-021-23497-y
5. Chen, Z., Zhang, Z., Qi, J., You, J., Ma, J., Chen, L.: Colorimetric detection of heavy metal ions with various chromogenic materials: strategies and applications. J. Hazard. Mater. **441**, 129889 (2023). https://doi.org/10.1016/j.jhazmat.2022.129889
6. Chen, D., Gao, K., Yang, J., Zhang, L.: Functionally graded porous structures: analyses, performances, and applications—a review. Thin-Walled Struct. (2023). https://doi.org/10.1016/j.tws.2023.111046
7. Farzin, M.A., Abdoos, H.: A critical review on quantum dots: from synthesis toward applications in electrochemical biosensors for determination of disease-related biomolecules. Talanta (2021). https://doi.org/10.1016/j.talanta.2020.121828

8. Geng, H., Vilms Pedersen, S., Ma, Y., Haghighi, T., Dai, H., Howes, P.D., Stevens, M.M.: Noble metal nanoparticle biosensors: from fundamental studies toward point-of-care diagnostics. Acc. Chem. Res. **55**(5), 593–604 (2022). https://doi.org/10.1021/acs.accounts.1c00598
9. Gupta, R., Verma, R., Kango, S., Constantin, A., Kharia, P., Saini, R., Chamoli, P.: A critical review on recent progress, open challenges, and applications of corrosion-resistant superhydrophobic coating. Mater. Today Commun. **34**, 105201 (2023). https://doi.org/10.1016/j.mtcomm.2022.105201
10. Harish, V., Ansari, M.M., Tewari, D., Yadav, A.B., Sharma, N., Bawarig, S., Barhoum, A.: Cutting-edge advances in tailoring size, shape, and functionality of nanoparticles and nanostructures: a review. J. Taiwan Inst. Chem. Eng. **149**, 105010 (2023). https://doi.org/10.1016/j.jtice.2023.105010
11. He, Y., Yin, Z., Wang, Z., Wang, H., Xiong, W., Song, B., Zeng, G.: Metal-organic frameworks as a good platform for the fabrication of multi-metal nanomaterials: design strategies, electrocatalytic applications and prospective. Adv. Coll. Interface. Sci. **304**, 102668 (2022). https://doi.org/10.1016/j.cis.2022.102668
12. Hopper, E.R., Boukouvala, C., Asselin, J., Biggins, J.S., Ringe, E.: Opportunities and challenges for alternative nanoplasmonic metals: magnesium and beyond. J. Phys. Chem. C **126**(26), 10630–10643 (2022). https://doi.org/10.1021/acs.jpcc.2c01944
13. Khurana, K., Jaggi, N.: Localized surface plasmonic properties of Au and Ag nanoparticles for sensors: a review. Plasmonics (2021). https://doi.org/10.1007/s11468-021-01381-1
14. Kumar, Y., Sinha, A.S.K., Nigam, K.D.P., Dwivedi, D., Sangwai, J.S.: Functionalized nanoparticles: tailoring properties through surface energetics and coordination chemistry for advanced biomedical applications. Nanoscale **15**(13), 6075–6104 (2023). https://doi.org/10.1039/d2nr07163k
15. Liu, Z., Liu, X., Ramakrishna, S.: Surface engineering of biomaterials in orthopedic and dental implants: strategies to improve osteointegration, bacteriostatic and bactericidal activities. Biotechnol. J.. J. (2021). https://doi.org/10.1002/biot.202000116
16. Liu, H., Fu, H., Sun, L., Lee, C., Yeatman, E.M.: Hybrid energy harvesting technology: from materials, structural design, system integration to applications. Renew. Sustain. Energy Rev. **137**, 110473 (2021). https://doi.org/10.1016/j.rser.2020.110473
17. Javaid, M., Haleem, A., Singh, R.P., Suman, R.: 5G technology for healthcare: features, serviceable pillars, and applications. Intell. Pharm. (2023). https://doi.org/10.1016/j.ipha.2023.04.001
18. Luppi, A.I., Rosas, F.E., Mediano, P.A., Menon, D.K., Stamatakis, E.A.: Information decomposition and the informational architecture of the brain. Trends Cogn. Sci.Cogn. Sci. (2024). https://doi.org/10.1016/j.tics.2023.11.005
19. Matzinger, M., Mayer, R.L., Mechtler, K.: Label-free single cell proteomics utilizing ultrafast LC and MS instrumentation: a valuable complementary technique to multiplexing. Proteomics (2023). https://doi.org/10.1002/pmic.202200162
20. Mehta, S., Suresh, A., Nayak, Y., Narayan, R., Nayak, U.Y.: Hybrid nanostructures: versatile systems for biomedical applications. Coord. Chem. Rev.. Chem. Rev. **460**, 214482 (2022). https://doi.org/10.1016/j.ccr.2022.214482
21. Miranda, R.R., Sampaio, I., Zucolotto, V.: Exploring silver nanoparticles for cancer therapy and diagnosis. Colloids Surf. B **210**, 112254 (2022). https://doi.org/10.1016/j.colsurfb.2021.112254
22. Mujahid, M.H., Upadhyay, T.K., Khan, F., Pandey, P., Park, M.N., Sharangi, A.B., Kim, B.: Metallic and metal oxide-derived nanohybrid as a tool for biomedical applications. Biomed. Pharmacother.Pharmacother. **155**, 113791 (2022). https://doi.org/10.1016/j.biopha.2022.113791
23. Nienhaus, K., Xue, Y., Shang, L., Nienhaus, G.U.: Protein adsorption onto nanomaterials engineered for theranostic applications. Nanotechnology (2022). https://doi.org/10.1088/1361-6528/ac5e6c
24. Nilghaz, A., Mousavi, S.M., Tian, J., Cao, R., Guijt, R.M., Wang, X.: Noble-metal nanoparticle-based colorimetric diagnostic assays for point-of-need applications. ACS Appl. Nano Mater. **4**(12), 12808–12824 (2021). https://doi.org/10.1021/acsanm.1c01545

25. Orooji, Y., Sohrabi, H., Hemmat, N., Oroojalian, F., Baradaran, B., Mokhtarzadeh, A., Karimi-Maleh, H.: An overview on SARS-CoV-2 (COVID-19) and other human coronaviruses and their detection capability via amplification assay, chemical sensing, biosensing, immunosensing, and clinical assays. Nano-micro Lett. **13**, 1–30 (2021). https://doi.org/10.1007/s40820-020-00533-y
26. Philip, A., Kumar, A.R.: The performance enhancement of surface Plasmon resonance optical sensors using nanomaterials: a review. Coord. Chem. Rev.. Chem. Rev. (2022). https://doi.org/10.1016/j.ccr.2022.214424
27. Pucci, C., Degl'Innocenti, A., Gümüş, M.B., Ciofani, G.: Superparamagnetic iron oxide nanoparticles for magnetic hyperthermia: recent advancements, molecular effects, and future directions in the omics era. Biomater. Sci. **10**(9), 2103–2121 (2022). https://doi.org/10.1039/d1bm01963e
28. Rej, S., Bisetto, M., Naldoni, A., Fornasiero, P.: Well-defined Cu_2O photocatalysts for solar fuels and chemicals. J. Mater. Chem. A **9**(10), 5915–5951 (2021). https://doi.org/10.1039/d0ta10181h
29. Skłodowski, K., Chmielewska-Deptuła, S.J., Piktel, E., Wolak, P., Wollny, T., Bucki, R.: Metallic nanosystems in the development of antimicrobial strategies with high antimicrobial activity and high biocompatibility. Int. J. Mol. Sci. **24**(3), 2104 (2023). https://doi.org/10.3390/ijms24032104
30. Sun, G., Chen, D., Zhu, G., Li, Q.: Lightweight hybrid materials and structures for energy absorption: a state-of-the-art review and outlook. Thin-Walled Struct. (2022). https://doi.org/10.1016/j.tws.2021.108760
31. Yang, H., Liu, J., Feng, X., Nie, F., Yang, G.: A novel copper-based metal-organic framework as a peroxidase-mimicking enzyme and its glucose chemiluminescence sensing application. Anal. Bioanal. Chem.Bioanal. Chem. **413**(17), 4407–4416 (2021). https://doi.org/10.1007/s00216-021-03394-5
32. Yang, J., Peng, Y., Li, S., Mu, J., Huang, Z., Ma, J., Jia, Q.: Metal nanocluster-based hybrid nanomaterials: fabrication and application. Coord. Chem. Rev.. Chem. Rev. **456**, 214391 (2022). https://doi.org/10.1016/j.ccr.2021.214391
33. Yang, S.M., Lv, S., Zhang, W., Cui, Y.: Microfluidic point-of-care (POC) devices in early diagnosis: a review of opportunities and challenges. Sensors (2022). https://doi.org/10.3390/s22041620
34. Yu, X., Wang, Y., Zhang, J., Liu, J., Wang, A., Ding, L.: Recent development of copper-based nanozymes for biomedical applications. Adv. Healthcare Mater. **13**(1), 2302023 (2024). https://doi.org/10.1002/adhm.202302023
35. Zhou, S., Yang, Y., Wu, Y., Liu, S.: Multiplexed profiling of biomarkers in extracellular vesicles for cancer diagnosis and therapy monitoring. Anal. Chim. ActaChim. Acta (2021). https://doi.org/10.1016/j.aca.2021.338633

Chapter 7
Applications of Metal Nanoparticles in Biosensing, Environmental Monitoring, and Medical Diagnostics

Abstract Metal nanoparticles (NPs) are important building blocks for colorimetric sensors, offering high sensitivity, rapid response, and wide application in biosensing, environmental monitoring and medical diagnostics. The specific properties of metal NPs, such as optical, electronic and catalytic properties, are reviewed to demonstrate their role in enhancing detection methods for various analytes like DNA, RNA, proteins, heavy metals and toxins. The integration of metal NPs into biosensors has led to significant improvements, resulting in more efficient detection of biomarkers, environmental pollutants, and pathogens. Key applications include the development of portable and cost-effective diagnostic devices, real time environmental monitoring systems, and food safety sensors. Moreover, the chapter highlights the role of hybrid and composite metal NPs in enhancing the sensitivity and selectivity of sensors. Metal nanoparticle-based colorimetric sensors are thus expected to transform diagnostics, environmental monitoring, and quality control within the biomedical and food industries, addressing critical global health and safety challenges.

7.1 Metal Nanoparticles in Biosensing

Due to their unique physicochemical properties (e.g., optical, electronic, and catalytic), metal nanoparticles (NPs) have attracted significant attention in the field of biosensing. These NPs, through application of advanced sensing techniques, efficiently identify various biomolecules, including DNA, RNA, proteins, bacteria, viruses, and metabolites such as glucose and lactose. Their incorporation significantly improves signal detection in sensing systems such as colorimetric methods, where visible changes correspond to alterations in nanoparticle arrangements (aggregation or dispersion) or shifts in extinction coefficients at specific wavelengths. Versatile processing methods employ noble metal NPs to create nanocomposites that can modulate energy transfer rates to the nanocrystals, thereby increasing or quenching the emission intensities from RGB NPs. A recently discovered plasmonic heating mechanism, which induces amplified gas emissions without the need for an external electrical field, further highlights the versatility of NPs.

A. C. Alguno et al., *Metal Nanoparticles-Based Nanoplatforms for Colorimetric Sensing*, Engineering Materials, https://doi.org/10.1007/978-981-96-6771-0_7

Currently, metal NPs are essential for both colorimetric sensing and enhancement-based applications, such as the analysis of biomarkers in sweat and bodily fluids. Advances in optical filters, hybrid materials, and related technologies have substantially improved sensitivity and specificity of detection compared to traditional fluorescence and immunochromatography methods. These developments have significant implications detecting NP–NP interactions with high sensitivity, as early-stage control over the assembly process enables the design of NPs with specific shapes and compositions, thereby minimizing interference and opening new avenues diagnostic tool development. Protein concentrations can also be measured using digital films that employ a tunable nanomaterials-layer exchange process combined with image processing techniques.

Additionally, metal NPs align well with sustainable, portable, and cost-effective applications, offering strong potential for clinical diagnostics and food quality monitoring. For instance, hybrid nanofibers containing copper NPs have been fabricated into touch-sensitive sensors exhibiting exceptional durability after repeated use. Moreover, the development of an implantable glucose nanoprobe based on silver NPs presents an innovative approach to improving diabetes patient management [18].

7.1.1 Detection of DNA, RNA, Proteins, and Other Biomarkers

This section highlights the impactful and innovative application of metal nanoparticles in colorimetric sensing. The following topics are addressed: (i) detection of DNA, RNA and proteins, which all key biomarkers; (ii) detection of glucose, melamine and urea as demonstrated in the studies of Tripathi & Bonilla-Cruz [20] and Zhao et al. [24]. In addition, this subsection discusses the future directions of the nanoplasmonic technology development. Given the broad scope of this field, and the impracticality of describing every characteristics in detail, this work presents representative studies in the later part of this section. It is expected that future studies will primarily focus on the development of metal nanoparticle-based nanoplatforms, and accordingly, many of the references cited herein are related to this specific class of nanoparticles.

The detection of biomolecules using metal nanoparticles (MNPs) as colorimetric sensors has attracted significant research interest in recent years. These biomolecules include DNA, RNA and proteins which serve as biomarkers for diagnosing diseases or monitoring human health [18]. The interaction between MNPs and biomolecules is primarily based on surface enhancement effects, wherein plasmonic energy is transferred from photons to electrons of MNPs, and then to the electrons in the biomolecules. These biomolecules can cause structural changes such as denatured, twisting or ruptured of the biomolecules, resulting in the alterations to the absorption spectrum, which may be visible to the naked eye. Several experimental methods have been developed to utilize MNPs as colorimetric sensors for biomolecule detection.

Overall, the results suggest that plasmonic metal nanoparticle measurements improve the capabilities of conventional bioanalytical techniques for detecting trace elements in complex natural samples.

7.2 Environmental Monitoring Using Colorimetric Sensors

Over the past few years, there has been an presence of excessive amounts of chemicals in water and soil, including heavy metals, pesticides and organic toxins, which pose serious threats to human health. As shown in Fig. 7.1, nanozyme sensor arrays for chemical detection are diverse. (a) Graphene-based nanozyme sensors doped with heteroatoms effectively detect pesticides. (b) H-GNs paper-based colorimetric sensors using heteroatoms-based graphene nanosheets exhibit specific reaction mechanisms for rapid sensing. (c) Pb^{2+} is detected by a graphene/Fe_3O_4-Au NPs sensing platform. (d) Smart GO/Au NPs interface enables colorimetric sensing of Oligonucleotide-based Hg^{2+}. These approaches improve sensitivity and selectivity while making devices portable for environmental monitoring and food safety applications.

Monitoring contaminants in water and soil in real time is highly desirable. Optical colorimetric sensor enhanced by metal nanoparticles are employed to achieve high

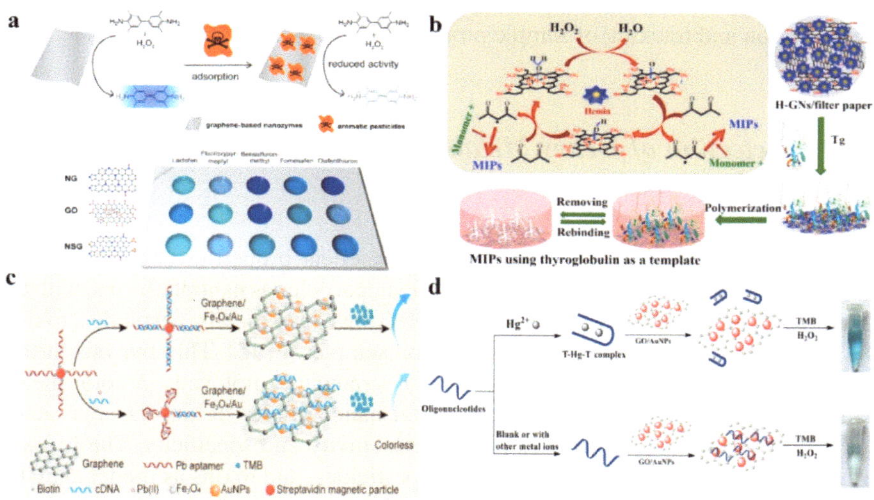

Fig. 7.1 **a** Nanozyme sensor arrays fabricated from heteroatom doped graphene for detecting pesticides. **b** The H–GNs paper-based colorimetric sensor was fabricated by the following principle. **c** The principle of the detection of Pb2+ on the graphene/Fe_3O_4-Au NPs sensing platform. **d** Schematic illustration of an oligonucleotide-based colorimetric sensor for Hg2+ detection using a smart interface of GO/Au NPs. Adapted from Wu et al. [22], Fig. 7.1. Licensed under CC BY 4.0 (https://creativecommons.org/licenses/by/4.0/)

sensitivity and resolution for the detection. This approach is suitable for real time detection with minimal or no sample processing, making it appealing device for environmental monitoring [7, 17]. In practice, just a few drops of the solution of interest are mixed with the substrate, enabling rapid detection that can be performed by non-specialists and is very fast. A discoloration reaction indicates the presence of contaminants on the surface, which can be visually observed with the naked eye. This provides means to monitor contaminants in water and soil in real time.

Metal NPs offer numerous advantages and can be used beneficially in the fabrication of colorimetric sensors. The oscillation of delocalized electrons in metal NPs, known as localized surface plasmon resonance (LSPR), depends on their morphology, size, and composition. These metal NPs absorb and scatter light at specific wavelengths corresponding to LSPR generating Strong electric near-fields that couple energy to adsorbed analytes, thereby enhancing their colorimetric signals. Furthermore, vibrating NPs produce local heat and increase that rises the temperature of physisorbed analytes, which can be measured through a change in color of the solution. Most notably, vibrating nanoparticles produce local heat that raises the temperature of physisorbed analytes, which can be measures by changing the color of the solution [8], allowing detection of targets of at femtomolar concentrations.

Recent advances in large-scale synthesis methods have enabled the incorporation of metal NPs into the sensing matrix. This has significantly enhanced colorimetric sensing performance, particularly for naked-eye detection at parts per billion levels. Furthermore this technique allows real-time color changes, facilitating rapid identification and tracking of sample properties whenever needed [6].

7.2.1 Detection of Heavy Metals, Pesticides, and Toxins in Water and Soil

This commendable analysis is based on the excellent properties and functions of colorimetric sensors constructed from metal nanoparticles as nanoplatforms, with the aim of applying them for the colorimetric detection heavy metals, pesticides, toxins, and other harmful substances in water and soil samples [3, 15]. This analysis further builds on the discussion of various detection strategies employees in colorimetric sensors, the majority of which rely on metal nanoparticles or similar approaches that utilize metal nanoparticles to enhance sensitivity and specificity. The focus is primarily on the detection of heavy metals, pesticides, and toxins as model analytes to demonstrate these mechanisms.

Moreover, the prospective directions in this swiftly evolving field have been proposed. Metal nanoparticles have been extensively developed as versatile nanoplatforms for colorimetric sensors through functionalization, modulation, assembly, and other innovative strategies undertaken in recent years. These advancements aim to improve sensitivity, selectivity and overall portability for practical applications in real-world scenarios. The discussion has not been limited to critical fields such as biomedicine, food safety and environmental monitoring but has been well incorporated in the analysis.

The detection mechanisms are described in detail and supporting case studies are presented to demonstrate their applicability. Various existing and emerging methods that rely on visible color changes of the analyte upon light exposure have been systematically developed as complex colorimetric sensors or assays. AuNPs, AgNPs, or other metal nanoparticles are mostly used as colorimetric reagents or mediators in most analyses [1]. The color changes observed during these detection processes which primarily indicate the presence of specific analytes, are best explained by phenomena such as the LSPR exhibited by metal nanoparticles, the complex aggregation behavior of the metal nanoparticles or the proximity effects between the nanoparticles and certain dyes, substrates or other chemical species [6].

7.3 Food Safety and Quality Control

This is an important global issue because foodborne illnesses and chemical contamination pose a significant threat to public health and economic stability worldwide [12]. Recent developments in nanotechnology have introduced new solutions, including nanosensors and nanomaterials, which are valuable in addressing these challenges. The application of nanotechnology in the food industry impacts safety of food, preservation, packaging, and quality. Nanotechnology helps extend product life by controlling gas and moisture levels, while nanosensors enhance food safety by detecting contaminants during processing and storage.

Nanosensors function based on the unique properties of nanomaterials which exhibit high sensitivity and can produce measurable signals in response to physical or chemical changes. These devices enable the real-time monitoring of parameters such as microbial contamination, chemical residues, spoilage and temperature. By employing these advanced technologies, food producers can ensure the highest standards of quality control and minimize on production losses. The development of lab-on-chip (LOC) devices that integrate nanosensors has also revolutionized food safety testing, allowing for the detection multiple contaminants at a single test and providing faster results than conventional methods.

Nanotechnology also plays a vital role in smart packaging systems which not analyze the freshness of perishable items but also help spoiled [10]. For example, packaging materials containing nanosensors can detect the volatile organic compounds that are produced during food decomposition. The features of nanosensors that include quick response, minimal sample size and compatibility with digital technologies, make them essential components of the modern food safety measures.

7.3.1 *Identification of Foodborne Pathogens and Chemical Residues*

Food safety is most often threatened by foodborne pathogens, including *E. coli*, *Salmonella*, and *Listeria monocytogenes*, as well as chemical residues like pesticides and veterinary drugs [4]. Detection has been substantially improved through the use of nanosensors based on nanoparticles, which offer high sensitive and specific. These devices work by transforming the material such as gold, silver, or carbon nanotubes, to produce optical, electrochemical or thermal signals to indicate the presence of contaminants [19]. The nanoscale dimension allows the detection of pathogenic organisms and other harmful substances long before they become visually apparent.

For example, optical nanosensors detect changes in optical absorption or fluorescence associated with pathogenic bacteria. Electrochemical nanosensors changes in electrical signal arising from the interaction between a bioreceptor, such as an antibody or DNA, and the analyte of interest [14]. Many bioreceptors are functionalized with nanoparticles to improve the binding specificity or increase the signal output. These rapid detection methods outperform conventional microbiological techniques, completing a tests in few hours compared to several days to traditional methods.

Nanosensor technology can also effectively monitors chemical residues, including pesticides and heavy metals. Surface enhanced Raman spectroscopy (SERS) uses nanoparticles to enhance chemical interaction signals, enabling detection of trace amounts of dangerous substances. Lab-on-a-chip (LOC) systems incorporating multiple nanosensors for a wide range of on-site testing have been proposed to provide a realistic approach for continuous monitoring of the food chain.

As described in earlier parts of this book, metallic nanoparticles and their aggregates have been employed for colorimetric sensing of various bioanalytes and chemical species [17]. These nanoparticles exhibit vivid colorful changes and high spectral sensitivity in aqueous environment due to their unique optical properties. Nanomaterial-based colorimetric methods have been effectively used to assess food quality and detect external chemicals such as veterinary drug residues.

Food safety challenges can be categorized could be broadly categorized into two group; microbial contamination and chemical residues in food products. Even small quantities of harmful microbes can pose dangerous, while products and packages containing pesticides, veterinary drugs, degraded plastics may also be harmful to consumers. Various types of nanomaterials are used as sensitive platforms to develop portable detection devices and instruments that use visible changes to monitor the food quality and chemical residues in agricultural produce.

Among the noble metallic nanoparticles, AuNPs are the most popular due to desirable properties such as having a reactive surface on which functional components, chemically inertness, particularly when compared to AgNPs [9]. Their nontoxic nature and suitability for instrumental analysis under high vacuum conditions make them especially useful for characterization and surface-enhanced technologies. Some of the most promising nanomaterial-based sensors used for ensuring the food safety and quality are presented together, along with their potential for further development of macroscale instruments and devices.

7.4 Medical Diagnostics and Point-of-Care Testing

Nanoparticles (NPs) are not only used in the analysis of various medical biomarkers including enzymes, proteins, drugs and metabolites but also in developing new methods of diagnosis [2]. The most popular approach is preparing spherical NPs because they are easy to synthesize and can be applied in various fields. In contrast, nanorods or nanoleaves with distinct shapes need additional time, effort and sometimes complex methods of synthesis procedure. Furthermore, the application of nanocomposite materials, which are materials composed of NPs dispersed in the matrix of another material, has recently emerged as an extremely sensitive method for detecting very low concentrations of analytes in diverse biological and chemical samples. As a general rule, NPs of various materials, shapes and sizes are engineered to possess specific and tunable properties suited to particular functional requirements. Normally, NPs are stabilized either through the adsorption of charged molecules or amino acids, or by the covalent attachment of polymer to their surfaces, allowing them to remain stable during the experiment procedures. This stabilization step is critical because non-specific species that remain unbound or are loosely bound to the nanoparticle surface must have washed away, as their presence can negatively affect the sensitivity and accuracy of the sensor. When the concentration of the target analyte in the solution is high, it can induce aggregation of the nanoparticles, which may reduce their sensitivity, On the other hand, if the NPs are too far apart, it may hinder the LSPR, and the shift in the color may be significant enough to indicate the presence of the analyte.

7.4.1 Cancer Markers, Infectious Diseases, and Genetic Disorders

Cancer is a complex group of diseases originated by genetic mutations and the abnormal growth of cells that can potentially spread throughout an organism. Genetic mutations such as single-nucleotide polymorphisms (SNPs) play an important role in cancer progression and can also increase the risk of developing this disease over time. However, most genetic SNPs are inherited from the parents, which on turn can be used as a strong cancer marker. For instance, the BRCA1 gene is a well-known cancer marker strongly related with breast and ovary cancer [16], but testing for it remains expensive for the majority of the population. The development of low-cost SNPs detection for field use can offer rapid and cost-efficient cancer screening.

Increasing efforts have been made to improve the portability of the devices for colorimetric SNP detection. The main strategy applied in these assays involves the interaction between immobilized oligonucleotide probe sequences on a solid substrate and the SNP target molecule in the liquid phase, which in turn bind the AuNPs [13]. This binding alters the LSPR of the AuNPs solution resulting in a color shift that can be visualized with the naked eye. Lasers and optical fibres have been developed to amplify these color changes in low resolution devices. For instance, a device combining a laser diode and a CMOS sensor has been fabricated to detect color changes after microarrays hybridization. Another portable device composed of a smartphone paired with a laser diode and a laboratory-on-a-chip containing the gold microarray. Alternatively, other devices such as Au films that utilizes spectral reflectance for detection have been designed and can be interpreted visually without additional equipment.

7.5 Integration with Modern Technologies

Metal nanoparticles based colorimetric sensors possess several advantages, including the following: (i) high stability and sensitivity; (ii) rapid response; (iii) favorable SPR properties; and (iv) simple preparation procedures. These characteristics make them attractive for wide range of applications. The underlying mechanisms rely on various methods such as the salt effect, the etching effect, the etching inhibition effect, and redox reactions. Novel colorimetric sensors can also be engineered to detect multiple properties in various sample type by properly selecting the reducing agents or substrates and by controlling specific experimental conditions. These properties may include pH, temperature and chemical concentration. The integration of metal nanoparticles based colorimetric sensors with emerging technologies holds great promise for expanding their application domains and may lead to the emergence of new research centered on metal nanoparticle based nanoplatforms. However, certain challenges must be addressed in order to optimize the performance of spray deposited metal nanoparticle-based sensors. In this context, metal nanoparticle based colorimetric sensing and electro-optically modulated technology plays a vital role [5]. For example, the shift in the LSPR spectrum of the deposited thin films of eight nanometer AuNPs is significantly influenced when exposed to air flow. In such case, the transmission across three distinct wavelengths is measured as a function of the angle of incidence. Coating the nanoparticles with polymers results in a significant suppression of the LSPR shift, thereby enhancing the control and stability of the biochip surface. This leads to a significant improvement in the uniformity of chemically conjugated NP films within the microfluidic platform. Finally, the LSPR shifts caused by the changes in the surrounding medium's permittivity are quantified and recorded for five commercially available and synthesized metal nanoparticle samples to measure their refractive index sensitivity [8, 11].

7.5.1 Smartphone-Based Sensors, Microfluidic Platforms, and Internet of Things (IoT)

Smartphone-based sensors, microfluidic platforms, and Internet of Things (IoT) have significantly advanced the development of colorimetric assays into the hand-held planar test strips. Cellular devices, spectrum analyzers, spectrophotometers, planar test strips, and paper-based microfluidic platforms are the major tools in this rapidly evolving field. Planar test strips and paper-based microfluidic platforms, also referred to as rapid detection tests, were originally developed for sensing analyte transformations and have been widely used to monitor the prevalence of diseases such as HIV/AIDS, Malaria, and as well as other infections and non-communicable diseases. The impregnation of absorbent filters or channels with glucose, protein, human chorionic gonadotropin (HCG), DNA, and RNA has gained industrial acceptance for rapid detection testing [21], primarily due to the integration of smartphone applications for results analysis and communications.

Colorimetric assays using suspensions of resonant metal nanoparticles have been performed on cellulose filters, sponges, tissues, and textile dispersions including cloth, wool, and thread for the detection of nutrients, pathogens, drug-related substances, pesticides, and biospecimens including glucose, protein, erythrocytes, HIV, and *E. coli* [23]. In these applications, the samples infiltrates fibrous channels or nanocomposites, which results in passive distribution and surface plasma resonance (SPR) or localized surface plasma resonance (LSPR) absorption. The paper tests display visual color changes, such as turquoise black or auburn brown which can be interpreted with the unaided human naked eye. However, standard cellular devices are not designed to capture and verify outcomes. Therefore, smartphones, spectrum analyzers or spectrophotometers are adapted quantify reflection, fluorescence, and absorption, which are fundamental to colorimetric lab-on-a-chip platforms. For high precision diagnostic profiling, enzyme-linked colorimetric analysis such as ELISA are still employed, requiring multiple washing and incubation steps. Simpler immunoassays, which could offer much less unmanageable immunoassays which could offer practical advantages, yet to be fully embraced, possible due to the dependence on specialized commercial equipment.

References

1. Ali, S., Chen, X., Shi, W., Huang, G., Yuan, L.M., Meng, L., Chen, X.: Recent advances in silver and gold nanoparticles-based colorimetric sensors for heavy metal ions detection: a review. Crit. Rev. Anal. Chem. **53**(3), 718–750 (2023). https://doi.org/10.1080/10408347.2021.1973886
2. Ahmad, A., Imran, M., Ahsan, H.: Biomarkers as biomedical bioindicators: approaches and techniques for the detection, analysis, and validation of novel biomarkers of diseases. Pharmaceutics (2023). https://doi.org/10.3390/pharmaceutics15061630

3. Ahmed, S.F., Kumar, P.S., Rozbu, M.R., Chowdhury, A.T., Nuzhat, S., Rafa, N., Mofijur, M.: Heavy metal toxicity, sources, and remediation techniques for contaminated water and soil. Environ. Technol. Innov.Innov. **25**, 102114 (2022). https://doi.org/10.1016/j.eti.2021.102114
4. Aworh, O.C.: Food safety issues in fresh produce supply chain with particular reference to sub-Saharan Africa. Food Control (2021). https://doi.org/10.1016/j.foodcont.2020.107737
5. Butt, M.A., Khonina, S.N., Kazanskiy, N.L.: Plasmonics: a necessity in the field of sensing-a review. Fiber Integr. Opt. Integr. Opt. (2021). https://doi.org/10.1080/01468030.2021.1902590
6. Burratti, L., Ciotta, E., De Matteis, F., Prosposito, P.: Metal nanostructures for environmental pollutant detection based on fluorescence (2021). https://doi.org/10.3390/nano11020276
7. Chen, Z., Zhang, Z., Qi, J., You, J., Ma, J., Chen, L.: Colorimetric detection of heavy metal ions with various chromogenic materials: strategies and applications. J. Hazard. Mater. **441**, 129889 (2023). https://doi.org/10.1016/j.jhazmat.2022.129889
8. Cui, Y., Zhao, J., Li, H.: Chromogenic mechanisms of colorimetric sensors based on gold nanoparticles (2023). https://doi.org/10.3390/bios13080801
9. Hammami, I., Alabdallah, N.M.: Gold nanoparticles: synthesis properties and applications. J. King Saud Univ.-Sci. (2021). https://doi.org/10.1016/j.jksus.2021.101560
10. Khan, S., Monteiro, J.K., Prasad, A., Filipe, C.D., Li, Y., Didar, T.F.: Material breakthroughs in smart food monitoring: intelligent packaging and on-site testing technologies for spoilage and contamination detection. Adv. Mater. **36**(1), 2300875 (2024). https://doi.org/10.1002/adma.202300875
11. Li, S., Qin, Z., Fu, J., Gao, Q.: Nanobiosensing based on electro-optically modulated. Technology (2023). https://doi.org/10.3390/nano13172400
12. Misiou, O., Koutsoumanis, K.: Climate change and its implications for food safety and spoilage. Trends Food Sci. Technol. (2022). https://doi.org/10.1016/j.tifs.2021.03.031
13. Mukhtar, M., Sargazi, S., Barani, M., Madry, H., Rahdar, A., Cucchiarini, M.: Application of nanotechnology for sensitive detection of low-abundance single-nucleotide variations in genomic DNA: a review. Nanomaterials **11**(6), 1384 (2021). https://doi.org/10.3390/nano11061384
14. Naresh, V., Lee, N.: A review on biosensors and recent development of nanostructured materials-enabled biosensors. Sensors (2021). https://doi.org/10.3390/s21041109
15. Rathi, B.S., Kumar, P.S., Vo, D.V.N.: Critical review on hazardous pollutants in water environment: occurrence, monitoring, fate, removal technologies and risk assessment. Sci. Total. Environ. (2021). https://doi.org/10.1016/j.scitotenv.2021.149134
16. Sanromán Iglesias, M., Grzelczak, M.: Using gold nanoparticles to detect single-nucleotide polymorphisms: toward liquid biopsy (2020). https://doi.org/10.3762/bjnano.11.20
17. Singh, H., Bamrah, A., Bhardwaj, S.K., Deep, A., Khatri, M., Brown, R.J., Kim, K.H.: Recent advances in the application of noble metal nanoparticles in colorimetric sensors for lead ions. Environ. Sci. Nano **8**(4), 863–889 (2021). https://doi.org/10.1039/D0EN00963F
18. Tavakkoli Yaraki, M. Nee Tan, Y.: Metal nanoparticles-enhanced biosensors: synthesis, design and applications in fluorescence enhancement and surface-enhanced Raman scattering (2020). https://doi.org/10.1002/asia.202000847
19. Thakur, A., Kumar, A.: Recent advances on rapid detection and remediation of environmental pollutants utilizing nanomaterials-based (bio) sensors. Sci. Total. Environ. (2022). https://doi.org/10.1016/j.scitotenv.2022.155219
20. Tripathi, A., Bonilla-Cruz, J.: Review on healthcare biosensing nanomaterials. ACS Appl. Nano Mater. (2023). https://doi.org/10.1021/acsanm.3c00941
21. Ulep, T.H. Yoon, J.Y.: Challenges in paper-based fluorogenic optical sensing with smartphones (2018). https://doi.org/10.1186/s40580-018-0146-1
22. Wu, Y., Feng, J., Hu, G., Zhang, E., Yu, H.-H.: Colorimetric sensors for chemical and biological sensing applications. Sensors **23**(5), 2749 (2023). https://doi.org/10.3390/s23052749

23. Zhang, X., Shi, Y., Chen, G., Wu, D., Wu, Y., Li, G.: CRISPR/Cas systems-inspired nano/biosensors for detecting infectious viruses and pathogenic bacteria. Small Methods (2022). https://doi.org/10.1002/smtd.202200794
24. Zhao, L., Li, L., Zhao, Y., Zhu, C., Yang, R., Fang, M., Luan, Y.: Aptamer-based point-of-care-testing for small molecule targets: from aptamers to aptasensors, devices and applications. TrAC Trends Anal. Chem. 117408 (2023). https://doi.org/10.1016/j.trac.2023.117408

Chapter 8
Challenges, Innovations, and Recommendations

Abstract Metal nanoparticles (MNPs) introduced in colorimetric sensing have gained considerable attention in various industrial applications, including food safety, environmental monitoring, medical diagnostics, due to their remarkable capabilities. This chapter reviews the current challenges, recent innovations, and future perspectives of MNP-based colorimetric sensors, with a focus on the detection of foodborne pathogen, chemical residues, and environmental toxins. Despite their advantages such as high sensitivity, rapid response, and ease of use, critical issues such as sensitivity, selectivity, reproducibility, and cost remain barriers to broader implementation. For this reason, novel synthesis approaches, including green synthesis methods and the development of bimetallic nanoclusters, are being explored to address these limitations and simultaneously enhance sensor performance. Moreover, the integration of artificial intelligence and machine learning is expected significantly advance sensor capabilities, enabling more precise detection and increased operational efficiency. This chapter outlines the recent development in MNP-based colorimetric sensing platforms and highlight their potential as miniaturized, real-time, portable detection systems for diverse applications.

The major food safety concerns are chemical residues and foodborne pathogens such as *E. coli, Salmonella* and *Listeria.* Recent developments in nanoparticle based nanosensors have improved detection methods by offering higher sensitivity and specificity [14]. These sensors, made from materials such as gold and carbon nanotubes, detect pathogens through optical, electrochemical or thermal signals, enabling the identification of contaminants before visible spoilage occurs [11]. Optical nanosensors indicate the presence of harmful bacteria through change in light absorption or fluorescence, while the electrochemical sensors detect variations in electrical signals caused by interaction of bioreceptors, such as antibodies or DNA, and the target analyte. These technologies provide rapid in detection, replacing conventional microbiological methods that typically require several days to complete. They also effectively measure chemical residues, for example, surface enhanced Raman spectroscopy (SERS) increases sensitivity for detecting

A. C. Alguno et al., *Metal Nanoparticles-Based Nanoplatforms for Colorimetric Sensing*, Engineering Materials, https://doi.org/10.1007/978-981-96-6771-0_8

low levels of toxic substances. On-site analysis is made possible with Lab-on-a-Chip systems that integrate several nanosensors, offering practical solution for food supply chain monitoring. In conclusion, integration of advanced detection technologies into real world applications through nanosensors helps reduce food-borne risks, thus contributing to consumer safety and confidence and promoting the growth of a sustainable food industry [18]. Colorimetric sensing platforms based on metal nanoparticles have been increasingly applied to monitor color changes in response to biochemical species, showing promising applicability in food and biological fluids analysis. However, current limitations include issues with sensitivity, selectivity, reusability, and affordability. There is the need to assess existing sensing technologies to support further innovation and development. This section is focuses on the application of metal nanoparticle based colorimetric sensing to address current challenges, analyze trends in technological development, and propose directions for further research. This is intended to support the scientific community, including researchers, developers and end users, in the advancement of colorimetric sensing platforms that are sensitive, specific, reusable and user friendly for the detection of biochemical and environmental targets.

8.1 Challenges in Metal Nanoparticle-Based Colorimetric Sensing

Metal nanoparticle (MNP) based colorimetric sensing systems suffer from several issues due to the proximate features of nanoparticles and the analytes. The main challenge is to ensure that the sensitivity and the selectivity are highly reliable. First, nanoparticles synthesized with fluctuation in size and distribution exhibit fluorescence and catalytic activity that directly affect detection precision [9, 15].

Another significant problem in this field is reproducibility. The random variation in nanoparticle production results in inconsistent output which undermines the credibility of these systems for real world applications. Previous studies have shown that parameters such as temperature, pH, and ionic strength influence nanoparticle stability and reaction rate [13]. These factors also affect the reproducibility of the experiments results.

A further challenge lies in integrating nanoparticles into practical devices. Most systems are designed under controlled laboratory conditions using minimal equipment and are not readily adaptable for use in remote or resource-limited environments. This difficulty is compounded by the absence of standardized protocols for nanoparticle synthesis and application, making it difficult for researchers to align their findings and collaboratively advance the field. Economic factors also pose limitations. The use of costly metals such as gold and platinum raises the overall production costs, making large scale applications economically unfeasible. Also, most existing synthesis methods are not yet scalable for industrial production, further restricting the broad application of these technologies.

To address these challenges, several strategies have been proposed. These include the development of alloyed or hybrid nanoparticles, and the integration of smart sensor algorithms to improve the synthesis and detection mechanisms [12]. Green chemistry approaches are also gain traction, as they reduce environmental footprint while improving economic viability. Future research should focus on developing synthesis methods that are scalable, reproducible, and cost-effective, and the MNPs are stable under various conditions [3]. Successfully addressing these issues will enable the broader adoption of MNP-based colorimetric sensing technologies in in medicine, environmental monitoring, and industrial control.

Metal nanoparticles possess unique and tunable optical properties making them promising candidates for chemical and biological colorimetric sensing. In recent years, more researchers have leveraged these properties for the development of sensors based on the change in color of metal nanoparticles, particularly noble metals such as gold (Au) and silver (Ag) [19]. While various reviews have been made on the mechanisms of colorimetric sensing using metal nanoparticles, mostly on general concepts and materials, leaving a gap in that explore the combined use of both gold and silver nanoparticles.

Furthermore, there is limited literature addressing the challenges, limitations, and future trends of metal nanoparticles in colorimetric sensor development. The main challenges include achieving optimal sensitivity, selectivity and reproducibility of the sensing response, managing the high costs of some metal nanoparticles, reducing environment and experimental interference, and the establishing standardized methods and techniques [5, 16].

The analysis of these challenges along with potential solutions and some directions for sensor development. Current research largely focuses on creating new materials and technologies for environmental monitoring and chemical detection. However, analyzing biological samples remains problematic due to the interference from other substance. One strategy to solve this problem is to increase the signal output. The use of metal nanoparticles in colorimetric sensing platforms has won great promise, primarily due to the LSPR effect, which result in a visible color shift depending on nanoparticle size and shape. This shift can be easily seen by the naked eye or through detection systems. Furthermore, bioactive molecules in the close proximity to metallic nanoparticles can also modify their LSPR response, expanding the range of detectable biological species. Gold nanoparticle based colorimetric nanoplatforms, in particular, have demonstrated high sensitivity in detecting a wide variety of analytes [1].

8.1.1 Sensitivity, Selectivity, Reproducibility, and Cost

The effectiveness of metal nanoparticle (MNP) based colorimetric sensing systems is best measured in terms of sensitivity, selectivity, reproducibility and cost. Sensitivity is vital since MNPs detect analytes at low concentrations by changing color. High sensitivity can be achieved by precise control of the size and composition of the nanoparticles, which is a technical challenge and a resource intensive process [17]. Selectivity is important in complex samples that contain several interferences. Researchers improve it by conjugating the nanoparticles with specific ligands, but this increases the complexity and the cost of the method due to the need for additional synthesis steps. Reproducibility is a major issue because even small changes in synthesis such as temperature can result in significant variations in the morphology and performance of the nanoparticles. This poses a serious problem in important application areas for example in medical diagnostics [2]. The issue of high cost is compounded by the use of noble metals such as gold because of their high cost is compounded by the use of noble metals such as golds, which are expensive and therefore limit the viability of large scale production. Copper or iron have been proposed, but they present concerns in terms of stability and performance. Some promising approaches for reducing the cost of synthesis while maintaining effectiveness include green synthesis methods that utilize plant extracts. Advances in computational modeling and machine learning may help improve control over the nanoparticle properties in order to enhance the reproducibility and effectiveness.

Future studies must aim to balance sensitivity, selectivity, reproducibility, and cost-effectiveness to transition MNP systems from experimental to practical applications. When these challenges are addressed, MNP-based colorimetric sensor can be leveraged to meet global needs. Colorimetric sensing, when combined with a biochemical recognition mechanism and the sensitivity of MNPs, offers visually detectable color changes across a wide range of applications. However, practical use reveals persistent challenges in sensitivity, selectivity, cost, and reproducibility that should be addressed intentionally. Sensitivity defines the lowest analyte concentration that ensure reliable results, by this is complicated by numerous factors that affect sensor performance. Selectivity remains critical, as it minimizes false outcomes in essential sectors such as in-vitro diagnostics.

Although new strategies for improving selectivity are under development, interference in complex environments continues to limit system reliability. Detection is made more difficult by biomarker concentrations, prompting the need for creative solutions in which strong signals outputs enable the analysis of trace targets [7]. There is also an urgent need for low-cost and user-friendly systems. However, designing such systems addressing the security associated with engineered nanomaterials that possess unique nanoscale features and intrinsic effects. As nanotechnology becomes increasingly integrated into products and processes, the potential health risks to human health and the environment concern remain a concern. This highlights the need for comprehensive knowledge and reliable data to inform regulation and assessments. The low reproducibility of MNP-based methods continues to undermine trust in these technologies. Standardized materials are essential for improving reproducibility, but current methods do not yet allow for adequate quantitative control over nanoparticle size. While complex social and economic factors make solution finding difficult, progress in production techniques has offer promission pathways forward. It is essential to maintain regulatory oversight and scientific rigor in this fast-changing evolving field [1, 6].

8.2 Recent Innovations and Emerging Technologies

Improvements have especially been made in the application of MNPs which have shifted the emphasis from earlier materials. These changes are directed towards enhancing the sensitivity, selectivity and the versatility of the sensors in order to fulfill the requirements for developing efficient and portable sensing devices [21]. For example, the catalytic activity and sensitivity of metal nanoparticle has been significantly improved by the DNA-templated synthesis of bimetallic clusters. DNA-templated Ag/Pt nanoclusters, for instance, have demonstrated excellent catalytic activity and can detect biomolecules such thrombin with a detection limit of 2.6 nM. These advancements have opened new research directions in the fields of biosensing and diagnostic applications.

The scope of MNPs is also expanding through their integration with other noble materials including perovskites. Perovskite-based sensors exhibit enzyme-mimicking activity and can therefore serve as an effective alternatives to traditional enzyme-based detection methods. Furthermore, core shell nanoparticles including Au@Pt has been reported to possess both plasmonic and catalytic properties, providing dual functionality for enhanced colorimetric response. A growing trend is the incorporation of MNPs into test strips and portable sensors. The use of MNPs as markers in lateral flow assays (LFAs) has greatly improved the detection of pathogens, allergens, and environmental toxins. These platforms are rapid, portable, and user-friendly, making them especially in resource-limited settings.

Recent developments in MNP-based colorimetric sensing have focused on increasing the efficiency, accessibility, and sustainability of detection systems. These advancements ensure their continued application in a variety fields including health-care, environmental monitoring and food safety. The development of highly sensitive and productive sensors and analytical tools has involved an interdisciplinary approach combining mainstream sciences, research methodologies, and emerging technologies [4]. This progress is supported by innovations in the synthesis of MNPs, which focus on novel synthesis methods, enhancement of their intrinsic properties, and integration with other techniques to obtain new properties.

8.2.1 New Synthesis Techniques for Metal Nanoparticles

Recent developments in MNP synthesis have been driven by advances in scalable and cost-effective methods with reduced environmental impact. Among these, green synthesis approaches are gaining increasing interest, using plant extracts, enzymes, or biopolymers as reducing and stabilizing agents [10, 20]. These approaches yield biocompatible nanoparticles and avoid harmful effects on the environment, making them suitable for biomedical applications.

Furthermore, novel chemical and physical synthesis approaches have been established. Solvothermal and microwave-assisted synthesis techniques provide good control over the size, shape, and surface characteristics of nanoparticles, each of which is crucial for colorimetric performance. Templates such as DNA and peptides enable the production of highly specific and functionalizable nanoparticles. The controlled synthesis of bimetallic nanoclusters, such as Ag/Pt, can be carried out using the DNA templates to regulate the catalytic activity, for example.

Hybrid nanoparticles are also the subject of intensive research. These nanoparticles composed of different metals or combined with other materials such as polymers and carbon-based nanostructures [14]. The synergistic effects of these hybrid nanoparticles led to enhanced catalytic activity and stability, resulting in a broader range of application in the field of colorimetric sensing. Nanoparticle production is increasingly being conducted using high-throughput and highly reproducible synthetic techniques such as additive manufacturing or flow chemistry. These approaches differ significantly from conventional methods that often produce larger quantities but lower-quality nanoparticles, which is a critical issue for commercial sensing devices.

The principles of colorimetric sensing using metal nanoparticles will be discussed. Likewise, detection mechanisms that would exhibit color changes or spectral changes upon interaction with analytes will be presented. The fundamental principles of colorimetric sensing with metal nanoparticles are described to clarify the reasons behind the vibrant color changes observed in colloidal metal nanoparticle systems when they interact with analytes or undergo changes in physical conditions. Color change is a simple and intuitive signal that can be readily understood by individuals regardless of their scientific background.

The color transition observed in metal nanoparticles is primarily attributed to alterations in surface plasmon resonance (SPR) associated with the aggregation and dissociation of the nanoparticles. The color changes are visible to the naked eye and can also be quantified using UV–vis spectroscopy. Although some studies have investigated the mechanisms behind these transitions, precise principles the color process whether influenced by chemical compositions or experimental setup, are not fully understood. The relationship between color change and plasmonic shift involved a unique and complex set of phenomena that nonetheless exhibit universal and consistent response to analytes or environmental changes such as salt concentration, pH, temperature, and refractive index. These transitions and the associated phenomena are highly valuable for the advancement of metal nanoparticles for colorimetric applications [6].

8.2.2 *Use of Artificial Intelligence (AI) and Machine Learning for Enhanced Detection*

New materials processing technologies, driven by advances in chemistry, material science, and bioengineering, have enabled the development of highly efficient sensing and imaging methods. However, the rapid and diverse nature of this technological progress has resulted in increased complexity, including challenges in design, fabrication, operation, and analysis. This level of complexity may limit the practical utilization of these technologies. In response, artificial intelligence (AI) techniques, particularly machine learning (ML), and computational methods have emerged as powerful tools that enhance detection, reduce false positives, and support the design of more effective sensors [8].

Large amount of data generated during the development and use of sensors directly beneficial for machine learning applications in colorimetric sensing. These algorithms can identify patterns in the data and optimize specific parameters to improve sensitivity. Moreover, they can predict sensor responses to various input parameters such as material composition, geometry, functionalization, or intended application. This interpretive ability supports both the experimental results and the adjustment of experimental parameters to enhance sensor performance. The integration of AI and ML not only increases the sensitivity of individual sensors but also contributes significantly to the overall enhancements of colorimetric detection technologies. There is a growing interest in sharing insights and case studies on how machine learning can be applied for the development, evaluation, or automation of colorimetric sensing. These exchanges are likely to inspire new concepts and foster cross-sector innovations [1].

References

1. Aldewachi, H., Chalati, T., Woodroofe, M.N., Bricklebank, N., Sharrack, B., Gardiner, P.H.: Gold nanoparticle-based colorimetric biosensors (2018). https://doi.org/10.1039/C7NR06367A
2. Ali, S., Chen, X., Shi, W., Huang, G., Yuan, L.M., Meng, L., Chen, X.: Recent advances in silver and gold nanoparticles-based colorimetric sensors for heavy metal ions detection: a review. Crit. Rev. Anal. Chem. **53**(3), 718–750 (2023). https://doi.org/10.1080/10408347.2021.1973886
3. Ali, A., Shah, T., Ullah, R., Zhou, P., Guo, M., Ovais, M., Rui, Y.: Review on recent progress in magnetic nanoparticles: synthesis, characterization, and diverse applications. Front. Chem. **9**, 629054 (2021). https://doi.org/10.3389/fchem.2021.629054
4. Bonnassieux, Y., Brabec, C.J., Cao, Y., Carmichael, T.B., Chabinyc, M.L., Cheng, K.T., Wu, Y.: The 2021 flexible and printed electronics roadmap. Flex. Printed Electron. **6**(2), 023001 (2021). https://doi.org/10.1088/2058-8585/abf986

5. Cheong, Y.H., Ge, L., Lisak, G.: Highly reproducible solid contact ion selective electrodes: emerging opportunities for potentiometry—a review. Anal. Chim. Acta (2021). https://doi.org/10.1016/j.aca.2021.338304
6. Cui, Y., Zhao, J., Li, H.: Chromogenic mechanisms of colorimetric sensors based on gold nanoparticles (2023). https://doi.org/10.3390/bios13080801
7. Cutshaw, G., Uthaman, S., Hassan, N., Kothadiya, S., Wen, X., Bardhan, R.: The emerging role of Raman spectroscopy as an omics approach for metabolic profiling and biomarker detection toward precision medicine. Chem. Rev. **123**(13), 8297–8346 (2023). https://doi.org/10.1021/acs.chemrev.2c00897
8. Feng, F., Ou, Z., Zhang, F., Chen, J., Huang, J., Wang, J., Zeng, J.: Artificial intelligence-assisted colorimetry for urine glucose detection towards enhanced sensitivity, accuracy, resolution, and anti-illuminating capability. Nano Res. **16**(10), 12084–12091 (2023). https://doi.org/10.1007/s12274-022-5311-5
9. Harish, V., Ansari, M.M., Tewari, D., Yadav, A.B., Sharma, N., Bawarig, S., Barhoum, A.: Cutting-edge advances in tailoring size, shape, and functionality of nanoparticles and nanostructures: a review. J. Taiwan Inst. Chem. Eng. **149**, 105010 (2023). https://doi.org/10.1016/j.jtice.2023.105010
10. Jeevanandam, J., Kiew, S.F., Boakye-Ansah, S., Lau, S.Y., Barhoum, A., Danquah, M.K., Rodrigues, J.: Green approaches for the synthesis of metal and metal oxide nanoparticles using microbial and plant extracts. Nanoscale **14**(7), 2534–2571 (2022). https://doi.org/10.1039/D1NR08144F
11. Kaya, H.O., Cetin, A.E., Azimzadeh, M., Topkaya, S.N.: Pathogen detection with electrochemical biosensors: advantages, challenges and future perspectives. J. Electroanal. Chem. **882**, 114989 (2021). https://doi.org/10.1016/j.jelechem.2021.114989
12. Konstantopoulos, G., Koumoulos, E.P., Charitidis, C.A.: Digital innovation enabled nanomaterial manufacturing; machine learning strategies and green perspectives. Nanomaterials (2022). https://doi.org/10.3390/nano12152646
13. Kumar, P., Thakur, N., Kumar, K., Kumar, S., Dutt, A., Thakur, V.K., Thakur, N.: Catalyzing innovation: exploring iron oxide nanoparticles-origins, advancements, and future application horizons. Coord. Chem. Rev. **507**, 215750 (2024). https://doi.org/10.1016/j.ccr.2024.215750
14. Fritea, L., Banica, F., Costea, T.O., Moldovan, L., Dobjanschi, L., Muresan, M., Cavalu, S.: Metal nanoparticles and carbon-based nanomaterials for improved performances of electrochemical (bio) sensors with biomedical applications. Materials **14**(21), 6319 (2021). https://doi.org/10.3390/ma14216319
15. Lee, S., Sim, K., Moon, S.Y., Choi, J., Jeon, Y., Nam, J.M., Park, S.J.: Controlled assembly of plasmonic nanoparticles: from static to dynamic nanostructures. Adv. Mater. **33**(46), 2007668 (2021). https://doi.org/10.1002/adma.202007668
16. Pham, T.C., Luong, C.M., Hoang, V.D., Doucet, A.: AI outperformed every dermatologist in dermoscopic melanoma diagnosis, using an optimized deep-CNN architecture with custom mini-batch logic and loss function. Sci. Rep. (2021). https://doi.org/10.1038/s41598-021-96707-8
17. Ramesh, M., Janani, R., Deepa, C., Rajeshkumar, L.: Nanotechnology-enabled biosensors: a review of fundamentals, design principles, materials, and applications. Biosensors (2022). https://doi.org/10.3390/bios13010040
18. Sharma, A., Ranjit, R., Kumar, N., Kumar, M., Giri, B.S.: Nanoparticles based nanosensors: principles and their applications in active packaging for food quality and safety detection. Biochem. Eng. J. **193**, 108861 (2023). https://doi.org/10.1016/j.bej.2023.108861
19. Singh, H., Bamrah, A., Bhardwaj, S.K., Deep, A., Khatri, M., Brown, R.J., Kim, K.H.: Recent advances in the application of noble metal nanoparticles in colorimetric sensors for lead ions. Environ. Sci. Nano **8**(4), 863–889 (2021). https://doi.org/10.1039/D0EN00963F

20. Soni, V., Raizada, P., Singh, P., Cuong, H.N., Rangabhashiyam, S., Saini, A., Nguyen, V.H.: Sustainable and green trends in using plant extracts for the synthesis of biogenic metal nanoparticles toward environmental and pharmaceutical advances: a review. Environ. Res. **202**, 111622 (2021). https://doi.org/10.1016/j.envres.2021.111622
21. Zheng, L., Cao, M., Du, Y., Liu, Q., Emran, M.Y., Kotb, A., Zhou, M.: Artificial enzyme innovations in electrochemical devices: advancing wearable and portable sensing technologies. Nanoscale **16**(1), 44–60 (2024). https://doi.org/10.1039/D3NR05728C

Chapter 9
Summary, Conclusions and Future Direction

Abstract This chapter comprehensively covers recent developments in the field of metal nanoparticle (MNP)-based colorimetric sensing platforms, with a focus on their applications for the detection of metal ions, biomolecules, and toxins. Various MNPs, such as AuNPs, AgNPs, PtNPs, and PdNPs, exhibit remarkable optical properties that enable localized surface plasmon resonance (LSPR) for rapid and sensitive detection with minimal instrumentation requirements. These colorimetric sensors offer significant advantages, including low detection limits, high selectivity, and the ability to provide visual and real-time monitoring. This chapter also explores the incorporation of MNPs with other advanced materials such as quantum-dots and hybrid nanostructures to enhance the performance parameters of the sensors. Despite the substantial progress achieved, several challenges remain, particularly in the areas of reproducibility, cost, and environmental stability. Future prospects indicate that substantial advancements can be realized through the development of smart sensors, hybrid nanoparticles and, the integration of artificial intelligence, potentially transforming colorimetric biosensing for biomedical, environmental, and industrial applications.

In summary, we have succinctly outlined the advances of a group of metal nanoparticles based on colorimetric nanoplatforms composed of AuNPs, AgNPs, PtNPs, PdNPs, Ag/Pd Huey NP, and MnNPs in colorimetric sensing across two main applications. Firstly, metal NPs-based colorimetric nanoplatforms can significantly enhance the development of colorimetric sensing for a wide range of important analytes, including metal ions, biomolecules, anions, or other materials, with excellent sensing performance [5, 10]. Their outstanding capabilities are demonstrated by their very low limit of detection and high selectivity compared to many other recently developed colorimetric methods. Secondly, these nanoplatforms also demonstrate notable advantages in enabling simple, rapid, and highly sensitive colorimetric sensing [9, 22]. They allow for easy and visual detection and exhibit several additional benefits.

Moreover, the mechanisms involved in colorimetric sensing using metal NPs including peroxidase mimetics, metal chaperone behavior, localized surface plasmon resonance effects related to shape, size, as well as the steric and entropy effects. The simplicity and broad applicability of these systems for both single-target and

A. C. Alguno et al., *Metal Nanoparticles-Based Nanoplatforms for Colorimetric Sensing*, Engineering Materials, https://doi.org/10.1007/978-981-96-6771-0_9

high-throughput screenings, along with their lower cost relative to other detection strategies, are further advantages. Looking forward, we anticipate that future breakthroughs in metal NPs-based colorimetric biosensing will result from integrating these metal nanoparticles with cell-free synthetic biology and targeting novel analytes. Additionally, applications that intersect with single-molecule studies and single nanoparticle plasmonic spectroscopy are expected to combine highly sensitive, accessible detection with deeper physical, chemical, or theoretical insights. The potential of such colorimetric biosensing platforms will go beyond advancing nanoparticles research and may contribute significantly to solving challenges in the field.

9.1 Summary of Key Findings

We have thoroughly explored a wide variety of metal nanoparticle-based nanoplatforms that have been used for colorimetric sensing in different areas of study including biomedical diagnostics, environmental monitoring, and chemical detection [11, 18, 26]. The use of these nanoparticles in advanced sensing technologies is facilitated, among other things, by their very small size, which enhances their interaction with light. In turn, the high surface-to-volume ratio of the NPs leads to increased optical changes compared to traditional systems, primarily due to the localized surface plasmon resonance (LSPR) effect that improves detection signals sensitivity. Moreover, the potential of enhancing colorimetric detection is further improved when these nanoparticles are combined with other advanced materials, such as CdSe quantum dots or different types of dyes.

A comprehensive investigation into the processes of synthesis, growth, and etching has been carried out to understand the sensing functionalities of metal nanoparticles [12, 21]. As deeper understanding of these mechanisms could hold the key to significantly improving photonic sensitivity, potentially transforming current detection methodologies. This chapter not only explains the fundamental scientific principles behind various colorimetric techniques but also highlights advanced colorimetric sensors and their potential applications. The role of these sensors is with growing relevance in the fields of biomedical diagnostics, environmental monitoring, and the development of more accurate, timely, and innovative detection systems through nanotechnology enabled sensing platforms.

We aim to suggest promising future directions that may guide the evolution of optical sensors in the coming years. It is widely recognized among researchers and scientists that modern investigations into real world applications are essential. These efforts underscore the need to address persistent challenges that have historically limited major advancements. Therefore, we have provided a comprehensive discussion of several potential avenues for future innovation, offering a sense of optimism about the opportunities these directions may present to the research and scientific community.

Looking ahead, we are hopeful that the advancements in metal nanoparticle-based nanoplatforms will remain a compelling subject within the academic and scientific communities. We strongly believe that these breakthroughs will lead to further improvement in the production of reliable colorimetric sensors applicable accross critical sectors including healthcare, environmental safety, and food quality monitoring. Ultimately, these developments may contribute to enhanced safety and improved quality of life for individuals and the communities.

9.1.1 Consolidation of Advancements in Metal Nanoparticle-Based Colorimetric Sensing

Recently, colorimetric sensors based on metal nanoparticles have attracted much interest because they can offer fast, sensitive, and relatively simple detection platforms [1, 2, 5, 13, 28]. In view of the fact that the present study is concerned with the development of guidelines for synthesizing and summarizing the progress made in this rapidly emerging field of metal nanoparticle based colorimetric sensing systems, it is imperative to compile a comprehensive set of guidelines that encompass all categories of metal nanoparticle based sensing systems, including those based on single component nanoparticles, bimetallic nanoparticles, and novel metal semiconductor hybrid structures. The latter employs a variety of sensing mechanisms; for example, particle plasmon resonance and shape transformation due to catalysis are both sufficient to cause noticeable changes in color. The combination of both aggregation and non aggregation approaches has enabled very simple colorimetric detection based on changes in particle color upon aggregation. These strategies enhance sensitivity and multiplexing by fully utilizing the optical properties of the nanoparticles used.

Recent developments in hybrid systems, which are based on various types of nanoparticles or incorporate additional materials such as graphene and enzymes, have significantly improved the selectivity and sensitivity of detection across a much wider range of analytes [16, 17, 23, 27]. Enzyme-functionalized nanoparticles have also shown great potential for detecting biochemical markers at the attomolar level. This sets new and impressive sensitivity benchmarks in various real-life applications. The stability of metal nanoparticles stability is also a significant concern. In this regard, the available data include the application of surface coatings with polymers and functional groups to increase the resistance of these nanoparticles in various environments. This improvement is essential because it enhances their performances in practical applications, where the conditions may differ greatly.

Furthermore, the development of advanced computational tools for data analysis has greatly improved detection efficiency and the adaptability of these systems various conditions. The emergence of portable and smartphone compatible sensors have further expanded the utility of these technologies in point-of-care diagnostics and environmental monitoring. These modern devices not only on the principles of metal nanoparticles but also enhance usability and accessibility, particularly in settings with limited or no laboratory facilities.

In future experimental work, it is recommended that the detectable range of analytes in colorimetric sensing systems be expanded. Furthermore, there is increasing interest in integrating of these advanced systems with digital technology to further boost their effectiveness. A significant opportunity also lies in the development of multifunctional nanoparticles for simultaneous diagnostics and potential therapeutic use, particularly in the field of theranostics. Metal nanoparticle-based colorimetric sensing will continue to be a vital and powerful approach for addressing critical global challenges in healthcare, environmental protection, and food safety, thereby playing an essential in the advancement of this promising technology.

9.2 Future Direction for Colorimetric Nanoplatforms

The changing nature of the colorimetric sensing platforms is indicative of a future defined by innovations aimed at addressing current system limitations and expanding their applications. Among the most commonly used in this area, metal nanoparticles, including gold and silver, will remain important due to their unique optical properties [1, 2, 5]. However, the rising interest in the application of copper and bimetallic nanoparticles suggests a shift to more affordable and environmentally friendly alternatives that do not compromise sensitivity and selectivity.

A major improvement is the integration of artificial intelligence and machine learning techniques which can enhance data analysis and pattern recognition in colorimetric analyses [14, 15]. This integration of the computational tools will expedite the processing of the complex samples and enable real-time diagnostics. Moreover, the construction of multifunctional systems based on various nanoparticles or functional materials, including graphene or polymers, is expected to improve the efficiency and the durability of colorimetric platforms.

In the point of care devices, the use of portable devices such as smartphone integrated colorimetric sensors is anticipated to grow [3, 20, 25]. These systems leverage the widespread use of smartphones to develop low-cost and user-friendly diagnostic tools for remote or resource-constrained environments. In addition, current advances in three-dimensional printing technology may transform the design of the colorimetric sensors, enabling the fabrication of compact, low-cost, single-use sensing architectures that can be integrated into test kits. Another emerging area is the development of multifunctional nanoparticles for both detection and treatment of diseases, paving the way for theranostics. These nanoparticles could identify disease biomarkers and deliver therapeutic agents directly to target sites, thereby supporting personalized medicine approaches.

Environmental sustainability will be a key consideration in the evolution of colorimetric nanoplatforms. Increasingly, researchers are focusing on green synthesis of nanoparticles and the use of recyclable or bio-degradable sensor materials to reduce environmental impact without sacrificing sensitivity or selectivity. These traits are particularly important in environmental monitoring of toxic wastes and in large-scale production. The establishment of regulatory standards will be essential as we transition towards advanced platforms that connect laboratory research to market-ready technologies. To achieve robust, reproducible, and scalable systems, collaborative among scientists, industry leaders, and policymakers is crucial. These developments will help solve global diagnostic and monitoring challenges by enabling the integration of high-performance, sustainable colorimetric sensors into real-world applications.

Thus, colorimetric sensing demonstrates remarkable versatility in designing highly sensitive and selective platforms for analytes analysis, particularly for point of care testing. The progress is seen in colorimetric lateral flow assays and the development of new labeling strategies and multiplexed tests supports this. For example, AuNPs of various shapes and plasmonic resonances can be deposited on urine or blood serum to detect prostate cancer-specific extracellular vesicles. Furthermore, nanosensors that are compatible with nucleic acids and other biomolecules will enable optimization of particle size, shape, surface chemistry, specificity, and sensitivity. Notable applications such as colorimetric sensors for pancreatic cancer diagnosis and paper-based biosensors with highly fluorescent gold nanoclusters illustrate this trend. Moreover, the use of well-defined host–guest interactions, such as functionalized calixarenes with AuNPs, may lead to the development of intelligent architectures for sensitive and selective colorimetric assays designed to detect biomarkers and monitoring of human inflammatory diseases, colorimetric detection of bacterial endotoxins.

9.2.1 Potential Impact on Analytical Chemistry, Diagnostics, and Real-World Applications

This colorimetric sensing phenomenon is entering a new era, marked by innovative applications across a wide range of fields, confirming its significant potential [4, 24]. This technology, which is both impressive and accessible, offers an intuitive and low-cost means to visualize and quantify a wide array of analytes that might be present in various solutions. Its functionality is particularly advantageous for real-time, on-site monitoring, which is not only fast but also precise, making it ideal for high-throughput detection applications that are increasingly common in today's demanding environment. The promise of simplicity, ease of use, and high sensitivity has greatly captured the attention of many chemists and researchers who seek to leverage this rapidly emerging technique. However, a major challenge that remains for researchers in the field is achieving accurate detection of diverse analytes. The development of colorimetric nanosensor presents a robust and responsive solutions to this fundamental issue. This sensors have exhibited remarkable growth alongside advancement in enabling technologies such as printable electronics, big data analytics, and the evolving Internet of Things. At the same time, these highly innovative nanosensors can serve as a fundamental components for smart devices that are customizable for numerous applications. These range from preventing natural product misidentification to enable the detection of single nucleotide polymorphisms using CRISPR system.

Beyond their sensing capabilities, nanosensors are distinguished by their vivid colorimetric properties and their suitability for multifaceted applications. These include use in bactericidal textiles, recyclable membranes for water purification, and efficient eco-friendly inkjet printing technologies that contribute to significant waste reduction. They have also been applied in practical areas such as bio-metallurgy, thermophotovoltaics, and the fabrication of structured chiral microenvironments that facilitate specific chemical reactions. Given the dynamic advancement in materials science and molecular chemistry, the integration of colorimetric nanoparticles is expected to expand across diverse fields. This evolution unlocks opportunities areas such as analytical chemistry, clinical diagnostics, and environmental monitoring. Consequently, these technologies are poised to indispensable tools in both scientific research and a wide range of industrial applications in the near future.

9.3 Final Thoughts on the Evolution of Colorimetric Sensing Techniques

The advancement of colorimetric sensing techniques represents a gradual transition from semi-quantitative to fully quantitative systems capable of addressing complex analytical problems. Initially, these methods relied on simple chemical reactions and visual estimation. Today, they are sophisticated systems that harness the optical properties of metal nanoparticles. A key contributor to this evolution is the use of LSPR, which has enabled the development of highly sensitive and selective colorimetric sensors [6, 7]. These platforms have been optimized by controlling the size, shape and surface chemistry of nanoparticles to serve diverse applications. The development of hybrid nanostructures and enzyme-conjugated nanoparticles has further enhanced colorimetric sensing capabilities.

The integration of digital and mobile technologies has significantly transformed the accessibility and utility of colorimetric sensing. Smartphone-based colorimetric sensors, in particular, have revolutionized diagnostic testing by making advanced analytical capabilities widely available and enabling accurate measurements with minimal technical expertise [8, 19]. This development highlights the importance of interdisciplinary collaboration in driving innovation. Future trends in the field are likely to shaped by sustainability and ethical considerations. The adoption of green synthesis methods and biodegradable materials in sensor construction will not only enhance sensor sensitivity but also promote environmentally sustainability and alignment with the sustainable development goals. By merging technological advancement with socially responsible innovation, these platforms will set new benchmarks for accessibility, accuracy, and eco-friendliness in next-generation diagnostic and monitoring systems.

Adaptability and continuous evolution are defining attributes of colorimetric sensing. Their enduring relevance across multiple disciplines demonstrates their capacity to meet emerging challenges, securing their role as an essential tools in contemporary science and industry. From early experiments observing color shifts in solutions to the creation of dual- or multimodal sensors, the trajectory of development points toward the use of low-cost and readily available materials such as polymers, paper, molecular sieves, and nanoparticles. Among the most studied materials, gold and silver nanoparticles remain of particular interest due to their unique optical properties derived from the localized surface plasmon resonance effect. As metal nanoparticle technology advance, these materials are regarded as leading candidates for future colorimetric sensing applications. Nonetheless, challenges remain. The synthesis of metal catalyst nanoparticles is often complex, which impede the broader adoption of nanoparticle based sensing technologies.

For practical implementation, at least four critical criteria must be satisfied: high sensitivity, high selectivity, low background noise, and suitability for point-of-care testing. The intrinsic sensitivity, selectivity, and visual detectability of metal nanoparticles-based colorimetric sensors, combined with their low operational costs, make them attractive for widespread use. In practical contexts, growing emphasis is placed on result accuracy and reproducibility, particularly for simultaneous detection of multiple analytes. Looking ahead, a critical challenge lies on the judicious selection of the appropriate signals to detect and generate, which will determine the future success and reliability of these sensing systems.

9.3.1 Anticipated Trends and Directions for Next-Generation Technologies

Anticipated trends and directions for next-generation sensing technologies promise exciting advancements, particularly in colorimetric sensing. While this approach holds immense potential for technological innovation, it remains largely in its early stages, with only limited number of solutions approaching commercial viability. For colorimetric detection to meet the growing demands for miniaturization, affordability, reliability, and rapid field-based measurements, substantial breakthroughs are needed to overcome existing limitations.

Significant knowledge has been acquired regarding the properties of nanoparticles and their in-situ formation within polymer matrices, leading to the development of various methods for controlled nanoprecipitation. Techniques such as precise temperature control and polymer grafting are effective strategies for adjusting nanoparticle size and dispersion, allowing fine-tuning of metal nanoparticles for targeted applications through modulation of polymer types and synthesis processes. Due to their instability, the large surface area of nanoparticles necessitates protective coatings.

Furthermore, the integration of specific receptors onto magnetic nanoparticles offers a reliable immobilization approach, with the nanoparticles acting as physical supports for the receptors.

In terms of practical applications, it is anticipated that colorimetric sensing techniques employing nanoparticles will see increasing applications across several fields, particularly in food safety diagnostics and in environmental monitoring. As new pathogenic strains continue to emerge in diverse environments, the adaptability of sensor recognition materials will be essential for improving detection performance. This adaptability enables the effective identification of various species and pathogens. In parallel with the development of both specific and broad-spectrum colorimetric detection assays, the implementation of efficient sample preconcentration and separation techniques will be vital for managing complex mixtures commonly encountered in real-world samples. The convergence of these advancements could significantly strengthen our capacity to safeguard health and ensure public safety.

For field applications, priority will be placed on the speed and reliability of color generation and detection. This highlights the potential of microfluidic systems to enhance the performance of portable analytical platforms. Regardless of which technologies ultimately prevail in the next generation of colorimetric detection systems, a key factor will be the seamless integration of nanotechnology into user-friendly formats that can be utilized by individuals with limited technical training. The scientific and engineering communities must remain proactive, continuously monitoring how emerging (bio)nanotechnologies, including innovations in colorimetry, contribute to improving safety and quality of life across multiple sectors.

References

1. Ali, S., Chen, X., Shi, W., Huang, G., Yuan, L.M., Meng, L., Chen, X.: Recent advances in silver and gold nanoparticles-based colorimetric sensors for heavy metal ions detection: a review. Crit. Rev. Anal. Chem. **53**(3), 718–750 (2023). https://doi.org/10.1080/10408347.2021.1973886
2. Ali, Z., Ullah, R., Tuzen, M., Ullah, S., Rahim, A., Saleh, T.A.: Colorimetric sensing of heavy metals on metal doped metal oxide nanocomposites: a review. Trends Environ. Anal. Chem. **37**, e00187 (2023). https://doi.org/10.1016/j.teac.2022.e00187
3. Bui, T.H., Thangavel, B., Sharipov, M., Chen, K., Shin, J.H.: Smartphone-based portable biochemical sensors: exploring recent advancements. Chemosensors (2023). https://doi.org/10.3390/chemosensors11090468
4. Chen, Z., Zhang, Z., Qi, J., You, J., Ma, J., Chen, L.: Colorimetric detection of heavy metal ions with various chromogenic materials: strategies and applications. J. Hazard. Mater. **441**, 129889 (2023). https://doi.org/10.1016/j.jhazmat.2022.129889
5. Chatterjee, S., Lou, X.Y., Liang, F., Yang, Y.W.: Surface-functionalized gold and silver nanoparticles for colorimetric and fluorescent sensing of metal ions and biomolecules. Coord. Chem. Rev. (2022). https://doi.org/10.1016/j.ccr.2022.214461
6. Duan, Q., Liu, Y., Chang, S., Chen, H., Chen, J.: Surface plasmonic sensors: sensing mechanism and recent applications. Sensors (2021). https://doi.org/10.3390/s21165262
7. Farooq, S., Zezell, D.M.: Advances in metallic-based localized surface plasmon sensors for enhanced tropical disease detection: a comprehensive review. Plasmonics (2024). https://doi.org/10.1007/s11468-023-02109-z

8. Horta-Velázquez, A., Ramos-Ortiz, G., Morales-Narváez, E.: The optimal color space enables advantageous smartphone-based colorimetric sensing. Biosens. Bioelectron. 117089 (2024). https://doi.org/10.1016/j.bios.2024.117089
9. Hu, X., Wang, P., Huang, C., Fang, C., Li, F., Ling, D.: The role of ligands on synthesis, functional control, and biomedical applications of near-infrared light-responsive metal-based nanoparticles. Coord. Chem. Rev. **502**, 215632 (2024). https://doi.org/10.1016/j.ccr.2023.215632
10. Jin, Z., Yim, W., Retout, M., Housel, E., Zhong, W., Zhou, J., Jokerst, J.V.: Colorimetric sensing for translational applications: from colorants to mechanisms. Chem. Soc. Rev. (2024). https://doi.org/10.1039/d4cs00328d
11. Jouha, J., Xiong, H.: DNAzyme-functionalized nanomaterials: recent preparation, current applications, and future challenges. Small (2021). https://doi.org/10.1002/smll.202105439
12. Karawdeniya, B.I., Damry, A.M., Murugappan, K., Manjunath, S., Bandara, Y.N.D., Jackson, C.J., Neshev, D.: Surface functionalization and texturing of optical metasurfaces for sensing applications. Chem. Rev. **122**(19), 14990–15030 (2022). https://doi.org/10.1021/acs.chemrev.1c00990
13. Krishnan, S.: Colorimetric visual sensors for point-of-needs testing. Sens. Actuators Rep. (2022). https://doi.org/10.1016/j.snr.2022.100078
14. Li, H., Xu, H., Li, Y., Li, X.: Application of artificial intelligence (AI)-enhanced biochemical sensing in molecular diagnosis and imaging analysis: advancing and challenges. TrAC, Trends Anal. Chem. (2024). https://doi.org/10.1016/j.trac.2024.117700
15. Low, J.S.Y., Thevarajah, T.M., Chang, S.W., Khor, S.M.: Based multiplexed colorimetric biosensing of cardiac and lipid biomarkers integrated with machine learning for accurate acute myocardial infarction early diagnosis and prognosis. Sens. Actuators, B Chem. **394**, 134403 (2023). https://doi.org/10.1016/j.snb.2023.134403
16. Montes-García, V., Squillaci, M.A., Diez-Castellnou, M., Ong, Q.K., Stellacci, F., Samori, P.: Chemical sensing with Au and Ag nanoparticles. Chem. Soc. Rev. **50**(2), 1269–1304 (2021). https://doi.org/10.1039/D0CS01112F
17. Montiel Schneider, M.G., Martín, M.J., Otarola, J., Vakarelska, E., Simeonov, V., Lassalle, V., Nedyalkova, M.: Biomedical applications of iron oxide nanoparticles: current insights progress and perspectives. Pharmaceutics **14**(1), 204 (2022). https://doi.org/10.3390/pharmaceutics14010204
18. Oliveira, B.B., Ferreira, D., Fernandes, A.R., Baptista, P.V.: Engineering gold nanoparticles for molecular diagnostics and biosensing. Wiley Interdisc. Rev.: Nanomed. Nanobiotechnol. **15**(1), e1836 (2023). https://doi.org/10.1002/wnan.1836
19. Qian, S., Cui, Y., Cai, Z., Li, L.: Applications of smartphone-based colorimetric biosensors. Biosens. Bioelectron.: X. (2022). https://doi.org/10.1016/j.biosx.2022.100173
20. Rasheed, S., Kanwal, T., Ahmad, N., Fatima, B., Najam-ul-Haq, M., Hussain, D.: Advances and challenges in portable optical biosensors for onsite detection and point-of-care diagnostics. TrAC Trends Anal. Chem. 117640 (2024). https://doi.org/10.1016/j.trac.2024.117640
21. Saleh, H.M., Hassan, A.I.: Synthesis and characterization of nanomaterials for application in cost-effective electrochemical devices. Sustainability (2023). https://doi.org/10.3390/su151410891
22. Tao, Y., Yi, K., Wang, H., Kim, H.W., Li, K., Zhu, X., Li, M.: CRISPR-Cas12a-regulated DNA adsorption and metallization on MXenes as enhanced enzyme mimics for sensitive colorimetric detection of hepatitis B virus DNA. J. Colloid Interface Sci. **613**, 406–414 (2022). https://doi.org/10.1016/j.jcis.2022.01.038
23. Truong, T.T., Mondal, S., Doan, V.H.M., Tak, S., Choi, J., Oh, H., Oh, J.: Precision-engineered metal and metal-oxide nanoparticles for biomedical imaging and healthcare applications. Adv. Colloid Interface Sci. 103263 (2024). https://doi.org/10.1016/j.cis.2024.103263
24. Umapathi, R., Sonwal, S., Lee, M.J., Rani, G.M., Lee, E.S., Jeon, T.J., Huh, Y.S.: Colorimetric based on-site sensing strategies for the rapid detection of pesticides in agricultural foods: new horizons, perspectives, and challenges. Coord. Chem. Rev. **446**, 214061 (2021). https://doi.org/10.1016/j.ccr.2021.214061

25. Xing, G., Ai, J., Wang, N., Pu, Q.: Recent progress of smartphone-assisted microfluidic sensors for point of care testing. TrAC, Trends Anal. Chem. (2022). https://doi.org/10.1016/j.trac.2022.116792
26. Yang, J., Wang, X., Sun, Y., Chen, B., Hu, F., Guo, C., Yang, T.: Recent advances in colorimetric sensors based on gold nanoparticles for pathogen detection. Biosensors (2022). https://doi.org/10.3390/bios13010029
27. Zhang, S., Rong, F., Guo, C., Duan, F., He, L., Wang, M., Du, M.: Metal–organic frameworks (MOFs) based electrochemical biosensors for early cancer diagnosis in vitro. Coord. Chem. Rev. **439**, 213948 (2021). https://doi.org/10.1016/j.ccr.2021.213948
28. Zhou, Y., Huang, X., Hu, X., Tong, W., Leng, Y., Xiong, Y.: Recent advances in colorimetry/fluorimetry-based dual-modal sensing technologies. Biosens. Bioelectron. **190**, 113386 (2021). https://doi.org/10.1016/j.bios.2021.113386

MIX
Papier aus verantwortungsvollen Quellen
Paper from responsible sources
FSC® C105338

If you have any concerns about our products,
you can contact us on
ProductSafety@springernature.com

In case Publisher is established outside the EU,
the EU authorized representative is:
Springer Nature Customer Service Center GmbH
Europaplatz 3, 69115 Heidelberg, Germany

Printed by Libri Plureos GmbH
in Hamburg, Germany